Statistics for Aquaculture

Statistics for Aquaculture

Ram C. Bhujel, PhD
Aquaculture and Aquatic Resources Management (AARM)
Asian Institute of Technology (AIT), Thailand

A John Wiley & Sons, Ltd., Publication

Edition first published 2008
© 2008 Wiley-Blackwell

Blackwell Publishing was acquired by John Wiley & Sons in February 2007. Blackwell's publishing program has been merged with Wiley's global Scientific, Technical, and Medical business to form Wiley-Blackwell.

Editorial Office
2121 State Avenue, Ames, Iowa 50014-8300, USA

For details of our global editorial offices, for customer services, and for information about how to apply for permission to reuse the copyright material in this book, please see our website at www.wiley.com/wiley-blackwell.

Library of Congress Cataloging-in-Publication Data

Bhujel, Ram C.
   Statistics for aquaculture / Ram C. Bhujel. – 1st ed.
     p. cm.
   Includes bibliographical references and index.
   ISBN-13: 978-0-8138-1587-9 (alk. paper)
   ISBN-10: 0-8138-1587-8 (alk. paper)
   1. Aquaculture–Statistics.   I. Title.
SH135.B48   2008
639.8072'7–dc22

                    2008036172

A catalogue record for this book is available from the U.S. Library of Congress.

Set in 10/12.5 pt Sabon by Aptara® Inc., New Delhi, India
Printed in Singapore by Markono Print Media Pte Ltd

1   2008

Dedication

Dedicated to my mother, wife, and children

Contents

Preface

Statistical background is essential for researchers in order to be able to design proper scientific experiments, analyze and interpret data correctly, and present their findings appropriately. However, most aquaculture scientists/researchers have limited knowledge in statistics.

Most agricultural universities offer statistics courses specific to agriculture or livestock husbandry. Surprisingly, no course on statistics specific to aquaculture exists in the world. Research in agriculture and livestock husbandry is relatively well-established, and several handbooks are available for these disciplines. Almost all aquaculture researchers have to find those books or find experts in their statistics departments for help in designing experiments and analyzing data. However most statisticians have only theoretical background and lack background in aquaculture or even biology. A statistician can't confidently suggest a design, analysis, and interpretation of data when it comes to the specific field situations. Most aquaculture researchers often feel helpless and face tough challenges at times of presentation and publication. They often end up without publication, which means huge funds and resources used for research are being wasted. At the same time, researchers quite often misinterpret the results and publish or present them with prejudice set in their mind. As a result, they are misled and misguide their readers. A simple erroneous conclusion and recommendation can have multiplier effect as it is cited by several others and the outcome is passed on to thousands of people, or even to millions. It is not practically possible to control the quality of all the papers, magazines, and newsletters by experienced scientists who have good statistical knowledge as well as background of the discipline. This has been a real challenge to the scientific community, especially in aquaculture research. Blue revolution is yet to come, and research in aquaculture is lagging behind agriculture and livestock. There are indications that the pace of aquaculture development has suffered quite a lot due to poor research and applicability of research results in the real industry. The cost of this must be huge, although it has not been estimated yet.

This handbook has been written as an attempt to mitigate the problems mentioned above, using the experience of postgraduate level teaching and working as a researcher in a regional hub of aquaculture development. One of the unique characteristics of the book is that it has actual cases as examples from real aquaculture research. Therefore, readers should get additional knowledge and practical problem-solving skills from this book. More importantly, it also covers nonparametric tests, realizing that they have become increasingly important and useful but are not covered by most other statistical books. Another aspect of this handbook is that advanced topics of covariance analysis, multivariate analysis, and cluster analysis have been described. Similarly, uses of computer software for complex designs have also been emphasized.

Most statistical books are loaded with statistical tables which may not have so much use in aquaculture research. Therefore, only a few statistical tables have been included in reduced forms.

It is, therefore, hoped that this book will serve as a toolkit for the researchers and educators involved especially in aquaculture and related fields.

Thank you.
The author
AIT, Bangkok

The Society Preface

The United States Aquaculture Society (USAS) is a chapter of the World Aquaculture Society (WAS), "a worldwide professional organization dedicated to the exchange of information and the networking among the diverse aquaculture constituencies interested in the advancement of the aquaculture industry, through the provision of services and professional development opportunities" (source: U.S. Aquaculture Society website: http://www.was.org/Usas/Default.htm). The mission of the USAS is to "provide a national forum for the exchange of timely information among aquaculture researchers, students and industry members in the United States. To accomplish this mission, the USAS will sponsor and convene workshops and meetings, foster educational opportunities and publish aquaculture-related materials important to U.S. aquaculture development.

The USAS membership is diverse, with 606 active members in 2008, representing researchers, students, commercial producers, academics, consultants, commercial support personnel, extension specialists, and other undesignated members. Member benefits are substantial and include issue awareness, a unified voice for addressing issues of importance to the U.S. Aquaculture Community, net-working opportunities, business contacts, employment services, discounts on publications and a semi-annual newsletter reported by regional editors and USAS members. Membership also provides opportunities for leadership and professional development through service as an elected officer or board member, chair of a working committee, or organizer of a Special Session or Workshop, special project, program or publication as well as recognition through three categories of career achievement (early career, distinguished service, and career). Student members are eligible for student awards and special accommodations at national meetings of the USAS, and have opportunities for leadership through committees, participation in Board activities, sponsorship of social mixers, networking at annual meetings and organization of special projects.

At its annual business meeting in New Orleans in January 2005, the USAS under the leadership of President LaDon Swann, voted to increase both the diversity and quality of publications for its members through a formal solicitation process for sponsored publications, including books, conference proceedings, fact sheets, pictorials, hatchery or production manuals, data compilations, and other materials that are important to US Aquaculture development and that will be of benefit to USAS members. Proper experimental design and analysis and interpretation and presentation of data are fundamental, yet challenging aspects of aquaculture research for most students, teachers and scientists. In this book project "Statistics for Aquaculture" Dr. Ram Bhujel draws from his experience as an aquaculture researcher and post-graduate level instructor to provide a practical reference book on statistical methods commonly used in aquaculture with examples from actual aquaculture research. Through collaboration with Blackwell Publishing on books projects such as these, the USAS Board aims to serve its membership by providing timely information through publications of the highest quality at a reasonable cost. The USAS thanks the author Dr. Ram Bhujel for sharing royalties which will help provide benefits and services to members and to the aquaculture community and Justin Jeffryes and Shelby Hayes (Wiley-Blackwell) for their cooperation. The USAS Publications Committee members include Drs. Wade O. Watanabe (Chair), Jeff Hinshaw and Jimmy Avery, with Ted Batterson and Rebecca Lochman serving as immediate past and current Presidents, respectively.

Wade O. Watanabe, Ph.D.
Publications Chair, United States Aquaculture Society
Research Professor and Aquaculture Program Coordinator
University of North Carolina Wilmington, Center for Marine Science
Wilmington, North Carolina USA

Acknowledgment

My heartfelt thanks are due to the United States Aquaculture Society for their supports and encouragement in publishing this book.

I also appreciate my students who have contributed in direct and indirect ways while preparing this book because this book is mainly based on the experience gained from teaching and working as a researcher at the Asian Institute of Technology. While teaching, I have also learned a lot from my students.

I would like to thank Mr. Chavan Balu, Mr. Truang Yen and Mr. Prabin Pury, who have assisted in converting some of the lecture notes into text, typing statistical tables, drawing some graphs, formatting, and file-managing tasks.

I would also like to thank my colleagues who have provided various types of support and encouragement while writing this book.

Chapter 1

Introduction

1.1 Background

Before learning statistics, one should know the scientific method. The ultimate goal of science is to understand and explain the natural and social phenomenon based on the conclusion of valid experiments and comprehensive observations. Observation and experimentation are the main two ways of generating "knowledge" about the natural world. In addition to observation and experimentation, scientific method also includes identification, description, and the theoretical explanation. In contrast, traditional knowledge is teachings and experiences passed on from generation to generation that is deeply rooted and developed as culture, customs, mythology, and language of the people as a way of living. Most traditional knowledge passes verbally from person to person across generations in the forms of stories, legends, folklore, rituals, songs, even regulation and laws. It refers sometimes to the matured traditions and practices in certain local communities that may differ from one community to another and may serve as a unique identity of particular communities. Most traditional knowledge is very valuable, but some needs to be tested in new contexts.

In many circumstances, observations and measurements are not possible; therefore, people have to imagine or hypothesize based on the limited available knowledge, which might not be true. Standard methods or procedures have been developed and are in use to carry out scientific inquiry or research. Depending on the nature of the research, outcomes can be broadly categorized into two groups: discovery and invention. Finding out things that already exist in the universe is called *discovery*, whereas creating or designing something new that never existed before is *invention*. Many scientific discoveries and inventions have played a significant role in changing the world and making human life a lot easier; for example, the invention of the bicycle, steam and jet engines, telephone, television, modern information technology, and so on. In agriculture, invention of new high-yielding varieties has brought about the green revolution, which is helping in feeding the ever-increasing population. Similarly, the development of high-milk-yielding dairy cattle has resulted in the white revolution. In aquaculture, we often talk about bringing about a blue revolution, but it has never taken place. People in this field are working hard in various ways to make it a reality.

A farmer grew five fish of a new strain in a single tank with excessive feeding. His fish grew 600 g in 3 months, and he compared the growth rate of his fish against the rates published in literature. He quickly noticed that the growth rate of his fish was almost double. He started advertising about his strain of fish around the world via e-mail, claiming that he had developed a new strain that grows two times faster than any other strains. Should we believe this claim? There is a similar story of a journalist who tested a new variety of rice, sowing a single seed in a well-managed plot to find a solution to the chronic shortage of rice in the Philippines in the 1950s. He harvested 1,000 grains from a single rice plant and, after extrapolating the yield, found 50 $t \cdot ha^{-1}$. Compared with $1 \cdot 2$ $t \cdot ha^{-1}$, the national average, he thought that, if that new variety was distributed to all the farmers, his country would not have any problems with rice shortages and could export and earn millions (Gomez and Gomez 1984). Is this a scientifically valid comparison and conclusion? The answer is "No." The scientific method involves a long process (Figure 1.1) that starts with observation then passes through all the way from imagination or hypothesis, designing and conducting experiment, data collection, analysis, and interpretation or reasoning. If the original imagination or hypothesis is proved, then it becomes a theory. A theory is not only a set of findings but also a set of well-developed themes and concepts that logically explain particular phenomena. Once a theory is widely accepted and applied, it becomes a universal law. Basically, a theory should be based on supporting data. This is often called a grounded or substantive theory, which is based on reality. It is derived from data gathered or generated systematically and analyzed through a research process. The process includes data collection from reliable sources or well-designed experiments, compilation of data, analysis and interpretation, presentation of findings, and theorizing or building a theory. Theory derived simply from phenomena is called formal theory, and if a theory is not empirically grounded in research, it is called speculative.

Scientific research is a long process and hard work. It can be boring but can also be exciting with the full joy of discovery or invention. Developing a career in research is climbing a steep ladder. Most researchers are either about to enter it or on the way to becoming researchers or great scientists, which means no one is 100% perfect. The knowledge and skills in doing research are reflected in their publications and presentations of research findings. The author has experienced several examples of lack of basic knowledge in statistics. For example, while attending a number of conferences and seminars, several researchers present their results, which say that fish fed with supplementary vitamin C or other nutrients had higher growth rate, survival, feed conversion ratio, meat yield, etc. than the fish fed from control treatment. However, these values were not statistically significant ($P > 0.05$). They even conclude and recommend that vitamin C or the other nutrients should be supplemented in the diet to increase the yield, which is wrong and misleading. Similarly, in survey type of research, researchers would

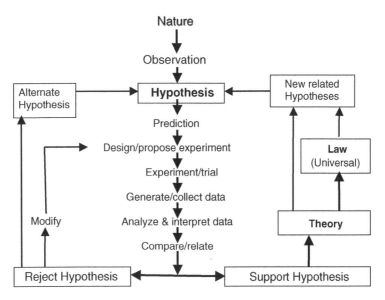

Figure 1.1 Scientific method.

say that farmers in district A had relatively bigger farms and higher fish production; however, these figures were not significantly different ($P > 0.05$) when compared using statistical tools. To claim this is wrong if the statistical analysis does not show any difference. Plenty of examples of this kind are found even in scientific literature, especially in aquaculture, which shows that there is a need for enhanced understanding of statistics among aquaculturists. This handbook has been written to help those researchers who are encountering problems, considering the fact that statistics is a must for researchers. However, commenting on the present status, Galton rightly said, "Some people hate the very name statistics, but I find them full of beauty and interest whenever they are not brutalized, but delicately handled by the higher methods, and are warily interpreted, their power of dealing with complicated phenomena is extraordinary."

1.2   History and definition of statistics

The word "statistics" originated from the Latin word "state," which means government. The states or the militaries were the first users of statistics and other advanced equipment or technologies, e.g. computers, remote sensing and geographical information systems, for the purpose of keeping records on the number of soldiers who died and returned alive during or after wars, the population of a city or state, and so on. Now, statistics is widely used by common people, e.g. football statistics, labor statistics, student enrollment statistics, and so on. As a plural noun, statistics means computed or estimated quantities, e.g. FAO statistics on production of rice (mt), aquaculture production (mt) of carps, catfish, tilapias, and so on. Statistic as a singular noun means a datum or numerical fact.

As statistics basically deals with numerical facts, it is considered a branch of applied mathematics. In fact, it is not only the mathematics; it is more about critical thinking and reasoning. Various scholars have tried to define statistics differently to reflect its processes and increasing roles. The simplest definition considers statistics a branch of mathematics that deals with the collection, analysis, and interpretation of data. The collection of data includes a good plan or design for a trial or questionnaire or survey, a clear procedure or method, materials or equipments to be used, and data compilation and storage. Data analysis means locating the central tendency, analyzing variability, exploring relationships or trends, and so on. The final part is the interpretation of the results and then making conclusions and recommendations. Therefore, the above definition was thought to be incomplete. Other definitions have been proposed. For example, statistics is the scientific study of numerical data based on variation in nature or the science of analyzing data and drawing conclusions, taking variation into account. This definition grasps the variation in data as the main characteristic. Variation or diversity is universal; For example, weights of fish vary, even if they are from the same age group, raised in the same cage, tank, or pond, and fed the same amount and type of feed. Even identical twins can differ in many attributes. If there is no variation in data, statistical analysis is not necessary. For example, Table 1.1 shows that variations (standard deviations) are zero in both sets of data in Trial 1. It can be seen that Treatment B resulted in higher survival of fish compared with Treatment A. But in Trial 2, replication 3 of Treatment A had higher survival than replication 4 of Treatment B. Due to this overlap, it is difficult to determine whether Treatment B results in higher survival of fish. Use of a statistical tool is not necessary in the case of Trial 1, whereas for Trial 2, a statistical tool is very important to make the right and confirmed decision.

Similarly, if all the fish spawned when a new hormone was used repeatedly, we could say that the new hormone is effective. But if only 90% of fish spawned among the fish injected with the new hormone, even in only one trial out of five, then statistics is necessary for making any decision. Variation in data means there are gray areas. In order to express the gray areas, researchers frequently use

Table 1.1 Hypothetical data showing with and without variation.

| | Survival of Fish (%) | | | |
| | Trial 1 | | Trial 2 | |
Replication	Treatment A	Treatment B	Treatment A	Treatment B
1	75	100	85	100
2	75	100	86	96
3	75	100	91	92
4	75	100	75	90
Mean	75.0	100.0	84.3	94.5
Standard deviation	0.0	0.0	6.7	4.4

the terms *almost*, *higher*, *lower*, *many*, *few*, or *relatively more*, and so on. But the results of the research supposedly conveyed by these words are not clear as these are vague and general words/statements. Those who have statistical knowledge would at least use percentage or probability. Statistical skill and knowledge therefore give individuals the skill or the power of interpretation and reaching conclusions. Furthermore, it teaches the techniques of presenting research results correctly and also helps in critical evaluation of literature published or planned to be published. Considering the various uses and importance of statistics, it has been defined as the science of decision making under uncertainty, as a body of methods and theory applied to numerical evidence in making decisions in the face of uncertainty, as a toolkit for problem solving, and so on.

Statistics is categorized as descriptive, which means use of data to report or describe the present status or the situation. It can be in either tabular or graphical form for the purpose of facilitating explanation. Selection of appropriate descriptive statistics is important. Another category of statistics is inferential, which means data are used to make inference, decisions, or conclusions based on the characteristics of the samples or parts of a whole. Knud–Hansen (1997) considers statistics as an inductive process where attempts to understand the whole are based on examining representative parts (or samples) through sampling and experimentation. Therefore, statistics can also be considered an art of collecting, presenting, describing, and interpreting data to understand our world and solve the problems.

There are several benefits of researchers having statistical knowledge and skill. According to Knud–Hansen (1997), it provides:

- skills of establishing and testing (proving/disproving) hypotheses
- knowledge about what and how much data to collect and not to collect
- confidence in results and interpretations
- power to critically review literature or others' work

In conclusion, statistics should not be considered only a branch of mathematics, but also an essential background for researchers, which ultimately becomes a way of their life. More importantly, it is a logical way of thinking that is necessary for everyone; therefore, according to H.G. Wells, "Statistical thinking will one day be as necessary for efficient citizenship as the ability to read and write."

1.3   Scope and application

1.3.1   In general

Attempts to define statistics have also been made based on its application. Statistics for biological sciences is often defined as "biometry," derived from two words: *bio* meaning life and *metron* meaning measure. In other words, it is the measurement of living organisms. The biological phenomena are so diverse and affected by many causal or environmental factors, and the factors themselves are variable, uncontrollable, and often unidentifiable; therefore, a fish pond is considered a black box! These are probabilistic in nature or statistical thinking, which means there is nothing absolutely right and nothing absolutely wrong! Some scholars express that it should be considered as a separate discipline. It is also referred to as "Bio-statistics," which means application of statistical methods to the solution of biological problems. Francis Galton, the cousin of Charles Darwin, has been considered the father of biometry. Other contributors and great scholars of biometry include Karl Pearson (1857–1936) and Ronald Fisher (1890–1962). Now, statistics has also been incorporated in various other disciplines, e.g. sociometrics, which means statistics combined with sociology. Similarly, when combined with economics, it is called econometrics; with psychology, psychometrics; with chemistry, chemometrics; and with forestry, forest biometrics. There are many more fields that are using statistics as an essential component in their disciplines. This clearly shows that the use of statistics exists in almost every field.

More importantly, this age is the era of information technology. The majority of organizations understand the value of data. They have maintained databases and stored a lot of data, even though they may not have used it yet. The number of such organizations is increasing daily. One organization may have several products and activities; all of them need to be recorded or maintained. Therefore, people who have skill in using the data are always in demand. The demand lies not only in maintaining the databases but also in analyzing them. Most organizations have started doing this, and they are used for making decisions or policies and formulating strategic plans. Large corporations have large volumes of data and require very skillful people to handle, analyze, and interpret that data. *Data mining* is the handling of such large volumes of data to explore, analyze, and discover meaningful patterns or trends so that forecasting is possible. Until now, most organizations have only stored data, but the time is coming for data mining to help policy making of these organizations. This shows that there will be a huge demand for statisticians in the near future.

1.3.2   In aquaculture

Although fish farming dates back about 4,000 years, FAO data show that its actual growth started only after the 1980s. It has now become the fastest growing food production sector. As it is the only alternative to compensate for the decline of capture fishery, it is expected to grow even faster to meet the 80 million mt (almost double the current level) production demand by the year 2050. However, currently, numerous challenges in this field have created an urgent need for more research within various disciplines, for example:

- increasing environmental problems caused by aquaculture development
- introduction of new aquaculture species causing threat to indigenous ones
- developing techniques of breeding, nursing, and culture of indigenous species
- increasing disease problems as a result of transboundary movement of aquatic species and intensification of culture systems

- development of low-cost feeds from locally available ingredients
- replacement for fish meal and fish oils, which are also used for livestock feeds
- economic studies for its viability or comparative studies with other sectors
- little is known about the roles of micronutrients and their interactions, e.g. minerals, vitamins, and fatty acids
- more studies on technology transfer or adaptive research and participatory on-farm trials
- food safety and quality

There is so much research to do for the full-fledged development of the aquaculture sector; however, most of the aquaculture scientists/researchers lack statistical knowledge and skills. Statistical background is essential for researchers to be able to design proper scientific experiments, analyze and interpret data correctly, and present them appropriately. Most aquaculture researchers have to find experts in their Statistics Departments or even outside to get help with designing experiments and analyzing and interpreting the data. They also face big challenges in publishing research articles. As a result, many of them often end up not disseminating the results after carrying out research, even when they have very fruitful findings; because of this, the whole aquaculture industry is suffering.

1.4 Questions

Q1. Why do science and statistics have such a close relationship?

Q2. Why do you think statistics is so important?

Q3. How will you apply statistical knowledge and skills in the future?

Q4. Debate whether one can or can't be a researcher without statistical knowledge.

Q5. Write an essay on the applications of statistics in aquaculture.

1.5 Practical exercise

Ex. 1. During the first practical session, instructors should guide students or trainees in developing basic skills of efficient spreadsheet data handling techniques. The following exercise would be useful:

- select, insert, and delete rows/columns
- insert and rename a worksheet
- enter numbers or texts
- create series of numbers, alphabets, dates, and their combinations
- perform data sorting and use of formula, functions, e.g. sum, etc.
- make good tables and different types of graphs

Data are given in Table 1.2 for practice.

Table 1.2 Batch weights of 15 fish from a trial at the Asian Institute of Technology, Thailand.

Treatments	Feeding Rate (%)	Replication	Batch Weight (g)	
			Stocking	Final Weight
A: Normal fish	1	1	663.3	784.2
(3 months old)	1	2	595.7	840.6
	1	3	581.6	814.9
	2	1	548.4	1005.1
	2	2	636.3	1304.8
	2	3	643.2	1259.1
	3	1	609.0	1513.9
	3	2	636.1	1432.4
	3	3	661.6	1291.3
B: Stunted fish	1	1	588.7	790.6
(12 months old)	1	2	493.5	734.8
	1	3	549.0	750.6
	2	1	505.5	1179.5
	2	2	517.8	1007.2
	2	3	549.3	1151.7
	3	1	560.2	1612.9
	3	2	526.6	1370.2
	3	3	572.8	1608.7

Note: Normal fish (Group A) were only 3 months of age, whereas stunted fish (Group B) were 12 months old but raised at high density under limited feeding conditions.

Chapter 2
Experimental units in aquaculture

2.1 Background

While planning and designing research, a clear understanding of experimental units and system is very important. Several factors, such as costs involved, purposes or objectives, availability of facility, applicability, management, climatic conditions, and experimental period, have to be considered. In this chapter, experimental systems commonly used for aquaculture research are briefly described.

2.2 Earth ponds

Earth ponds are the most common and cheapest aquaculture systems in most countries and locations. Therefore, research results of ponds trials are directly applicable and have wide use. The sizes of ponds used vary depending on the purpose of culture and availability of land. A nursery pond may range from 50 to 200 m^2, whereas a grow-out pond may vary from 500 m^2 to 1 ha, or even bigger. However, ponds as experimental units for research purposes should not be large, as land itself is expensive and research requiring adequate replications is costly. Normally, 50–100 m^2 for nursing and 200–400 m^2 for grow-out trials should be adequate. Unlike a land plot, where fertility of soil varies greatly, environmental conditions within a pond are very uniform if all ponds are supplied with the same water, as nutrients or chemicals can disperse freely in the water. However, some environmental parameters that are directly or indirectly associated with the survival, growth, and reproductive performances of fish among ponds may differ. For example, if a pond is close to an irrigation canal, road, or shade, gradient toward the pond's other side (see Section 7.3 for more information) may exist and need blocking (Figure 2.1). Bamboo, wooden, and/or metal walkways are built in experimental ponds in order to facilitate sampling of

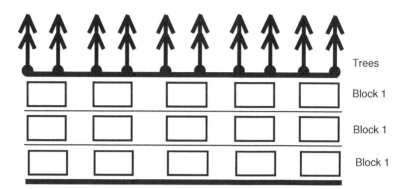

Figure 2.1 Earth ponds in a series next to windbreaker trees.

fish and water. While conducting an experiment in a pond system, the following should be considered:

- Birds can be a problem because they can eat experimental fish, and we can't expect the same level of predation in all units. Regular patrol by security guards, cover provided by nets, plastics hanging on ropes tied across the pond, and various other means would mitigate this problem to some extent.
- There is high chance of predatory fish and other animals, e.g. snakes, entering from outside, especially during a rainy/flood season. Double screens in inlets or outlets, net enclosure fences, sufficiently high dykes, and monitoring of water level may avoid the problem.
- During a rainy season, especially in low-lying areas, flooding can be a problem and could sweep away all of the experimental fish. In such areas, proper scheduling is needed or arrangement for flood monitoring and control is necessary if avoiding the flood schedule is not possible.

2.3   Hapas and cages in ponds

It is quite difficult to design or conduct research using ponds as replicates unless there is a large facility or specially designed smaller ponds are available. In some cases, it can be prohibitively expensive. However, a pond can be split into several experimental units by setting up hapas or cages, or hapas can be installed in rows (Figure 2.2a, b). There can be gradients from the edge of the pond to the center, as there are more chances for noises and shallowness at the edge compared with the central parts of the pond. If the hapas or cages are arranged in rows, the rows should be considered as blocks so that variability can be separated while carrying out the statistical analysis. As the nutrients/chemicals have no free movement from one pond to another, there is a good possibility that one pond may differ from another. Therefore, if more than one pond is used for the same trial, each pond should be considered a block, which means that each treatment should be allocated in each pond randomly. Nevertheless, in such a case, ponds or the blocks serve as replicates.

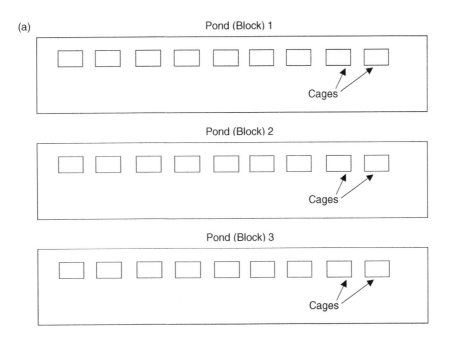

Figure 2.2 (a) Cages in three ponds. (b) An experimental pond with cages.

Hapas or cages have advantages for experimental purposes compared with ponds without because problems of predation by birds can be easily avoided by covering them. Also, cages and hapas can be lifted if the water level increases to avoid loss of fish due to flooding. However, cages and hapas can be expensive and setup time-consuming. Although sampling and harvesting of fish is easier, hapas or cages make it easier to steal fish, if there is a theft problem.

2.4   Cages in lakes or reservoirs

In many locations, lakes or reservoirs may serve as the experimental facility. Due to their large size, many experimental units can be set up in the same water; therefore, this is suitable for a completely randomized design (see Section 7.2). There is a chance of low variability or higher chance for any treatment to show its real impacts. However, care should be taken if cages are at or near the bank of lakes or reservoirs, where sunlight or shade can affect the rows at or near the bank more than others. In this case, cages can be arranged in rows along the bank, the rows can be considered blocks (Figure 2.3a, b). Randomized complete block design would be suitable (see Section 7.3). At the same time, if the experimental location is close to the water source or human activities, such as boating, additional blocking is necessary, which means two-way blocking.

2.5   Tanks

Conducting research in tanks allows more control over environmental factors and is also easier than research in ponds with cages, hapas, or pens. Various types of tanks can be built or chosen, e.g. circular, square, or rectangular. However, building tanks can be costly and care should be taken while designing a trial; a small mistake can ruin the research, requiring repetition. If tanks are supplied with the same source of water recirculated (bio- or mechanically filtered), then variability in water quality parameters other than temperature would be very low. On the other hand, outdoor tanks without water recirculation/exchange vary greatly. Tanks close to walls and water supply canals (Figure 2.4a, b) can affect water quality parameters and thereby fish growth and survival. Therefore, water quality parameters of each tank need to be monitored or measured. As tanks are smaller in size, productivity of the system or yield is expressed or compared normally in terms of $kg \cdot m^{-2}$ or $kg \cdot m^{-3}$.

2.6   Aquaria

Aquaria are the best experimental units for small-scale experiments, especially in a laboratory or fish hatchery under shade. These are easy to set up and handle. Researchers can have a large number of experimental units in most cases; therefore, they won't be constrained by the facility, shortage of replication, or

(a)

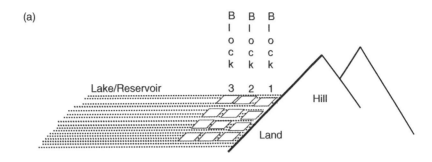

(b)

Figure 2.3 (a) Arrangement of experimental cages in lake. (b) Cages in a lake designed for a research station.

treatment levels. They can even be stacked to one another (Figure 2.5a, b). The height can be a block in such a case. A researcher can maintain better control over the environmental factors, which means there is less chance of masking the real effects of treatments. Most breeding and some feeding research is done using aquaria. However, the results of the trials in aquaria may have limited application on a commercial scale.

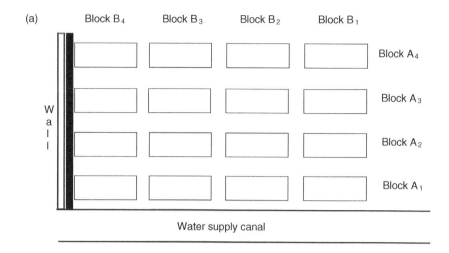

Figure 2.4 (a) Experimental tanks. (b) Tanks constructed for a research station.

2.7   Farmer's field: participatory research

Any research conducted within research stations or university facilities is called *on-station research*, which is directly controlled by people who are well aware of the research objectives or purposes. As the ultimate purpose of the on-station research is to serve a larger population, research outcomes will have to be tested in the field situation where there is less control of factors. Before making any policy recommendations, they have to be piloted and determined whether it works for the real beneficiaries. On the other hand, government extension organizations can conduct pilot studies using the facilities of selected farmers who are able to follow a set of guidelines. The research trials conducted with farmers – the ultimate beneficiaries – using their facilities is called *participatory research*. There is a popular saying, "farmers are researchers." By doing this, farmers also learn how to design a simple trial for comparisons when there is a need in later stages. At the same time, most extension workers think their responsibilities are only to transfer the research outcomes and the well-developed technology packages to the farmers. But, as no technology is perfect, they need continuous improvement and adaptation to special local conditions. Therefore, plenty of room for research

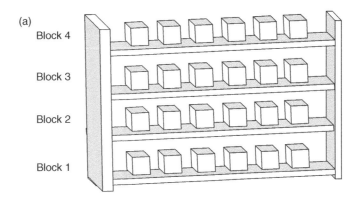

(a)

Block 4

Block 3

Block 2

Block 1

(b)

Figure 2.5 (a) Arrangement of aquaria for research. (b) Large buckets used for a trial.

is always there. An example of how a two-way block design can be adjusted for participatory research is shown in Figure 2.6a, b; there are four regions, and three agroecological zones are stretched in each region.

Although control of factors is limited, research outcomes can be extremely valuable because they are directly drawn from exactly the same conditions and thus are directly applicable. Participatory research is quite difficult to manage because suitable farmers might be far away from each other, and the researcher also has to deal with many types of people who may have a wide range of cultural backgrounds, beliefs, education, and/or income. Sometimes it can be very costly even though farmers' facilities are available for use free of cost because they may need to be adjusted or developed to meet the requirements of the research.

At the same time, the survey research can also be designed considering various factors, e.g. locations, elevations (low or high), soil types (fertile or unfertile), area (irrigated or nonirrigated), and so on. For example, in Figure 2.7, distance

(a)

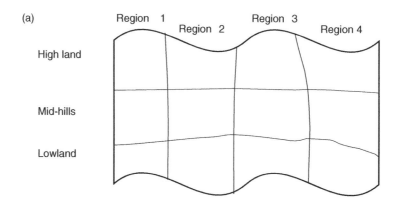

(b)

Figure 2.6 (a) Two-way blocking (4 × 3 design, see Section 7.2 for details). (b) One of the farmer's ponds used for participatory research and demonstration program.

from the sea (Factor A) and elevation (Factor B) can be considered two-way factors affecting the productivity of ponds in the area of shrimp culture.

2.8   Conclusion

A researcher should be very clear about what experimental unit is to be used. Individual fish within the same aquarium, tank, cage, hapa, or pond can be experimental units or replicates if all the fish are tagged. Conducting a breeding trial with hormone injection to determine the hormone efficacy in producing eggs is an example. In such a case, either number or volume of eggs is counted

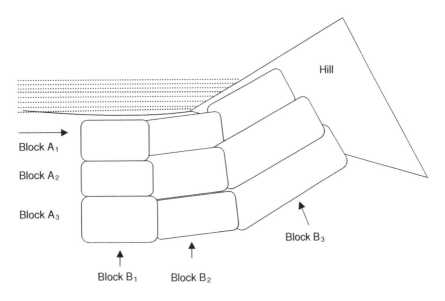

Figure 2.7 Two-way blocking (Factor A, sea; Factor B, elevation).

or measured per female or per spawning, which can be converted into relative fecundity or per kilogram of female.

Selection of a culture system for research depends on the researcher. Unless there is specific need, aquaria or tanks should not be selected simply because they are easier to manage and cheaper. The outcome of the research can have limited application in the real field situation. On the other hand, research in larger experimental units is unnecessary if they are unmanageable. Sometimes, it turns out to be waste of time, funds, and efforts if the potential factors cannot be controlled or monitored properly. Use of relatively smaller experimental units and a well-planned and properly managed experiment can generate adequate data. Appropriate statistical analysis can provide sufficient information to make precise and useful inferences.

2.9   Questions

Q1. What is an experimental unit?
Q2. What do you mean by experimental subunit?
Q3. Under which conditions can tanks be used as experimental units? And under which conditions can individual fish of the tanks be used?

2.10   Practical exercises

Ex. 1. If a researcher is planning to conduct an experiment to determine the effects of four different types of hormones in spawning, what should be the experimental unit and which experimental system should be used? How many of the experimental unit and the system would you suggest using?

Ex. 2. You are planning to conduct an experiment in earthen ponds to compare two strains of tilapia for breeding performance. How could you use these ponds for your trial? Estimate the number of each strain of fish needed based on different plans.

Chapter 3

Sampling and data collection

3.1   Sampling principles and methods

The entire collection of all possible organisms, objects, or observations of a specific characteristic of interest in a certain geographical area is *population*. In some research trials, measurement of all organisms may be possible, but most of the time, monitoring of an entire population is difficult, costly, impractical, and often impossible. For example, measuring of all 50 or up to 100 fish individually in a few experimental tanks or aquaria is possible. But if the tanks are larger, having 200 fish each or more, obtaining individual measurements of length and weight is tedious. Moreover, if we are supposed to find out the average weight of a fish species of a reservoir or lake, it is impossible to catch all the fish of that species. Therefore, instead of monitoring or measuring all of the organisms, manageable portions of the population are taken as samples. There is no hard and fast rule about the size of the sample. It varies widely, e.g. 1%, 5%, 10%, 25%, 50%, and even more depending on the situation. The larger the sample, the higher the accuracy and reliability will be, but cost increases with the sample size. Therefore, there is a trade-off between cost and accuracy. The main point is that the sample should be minimal but should be taken randomly so that all members of the population have an equal chance of being selected and the sample will be representative of the whole population. In order to make the samples representative, appropriate sampling techniques or methods have to be followed. The main methods of sampling are as follows:

- **Random sampling:** the common method by which samples are randomly chosen at a single stage from a population. For example, selecting 100 fish randomly, without any prejudice for size, color, or any other characteristics, from a tank containing 500 fish to measure individual weight and length. Random sampling needs to be done from each replicate, group, or strata of the whole experimental system.

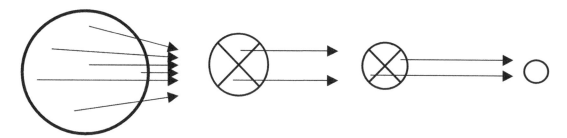

Figure 3.1 Three-stage sampling.

- **Systematic sampling:** sampling that is carried out at certain intervals of time. For example, monitoring water temperature and dissolved oxygen at 6 a.m. and 3 p.m. when these are minimum and maximum, respectively, in order to catch the window of critical times.
- **Stratified sampling:** the representative method of sampling from all the strata or groups. For example, a sample from each size class, e.g. large, medium, and small size groups of fish, to represent the whole population of differential size. The factor group can be included as block while analyzing the data.
- **Cluster sampling:** involves selection of certain groups or clusters first in order to avoid the types that are not necessary to include. Then, random selection within each cluster is done. For example, selecting 10 representative provinces of Thailand out of 76 to study the aquaculture development, followed by the selection of 35 farmers randomly from each province for interview.
- **Multistage sampling:** when the population is too large, sampling should be done a few times instead of just once. Figure 3.1 shows that first sampling is random. At the second stage, samples can be taken from two opposite sides (up and down) out of four quarters of a circle. Then at the third stage, again, selecting opposite sides (left and right) could produce a good sample.

3.2   Parameters or variables

Characteristics of populations are called *parameters*. The word "parameter" is the combined form of Greek words "*para*" and "*metron*," which mean beyond measure. An example is the population or true mean of 30,000 fish grown in a pond of 1 ha size, which is almost impossible to find by measuring individually. The true or population mean is a single value at a point of time, and it does not vary but is almost impossible to find in most cases. Parameters have to be estimated by sampling and may vary with the method of sampling and other factors. The word "variable" is popularly used; it reflects the properties with respect to which individual organisms or objects vary in some ascertainable way due to some cause(s). Any scientific inquiry or research begins with thinking about "cause and effect" relationship. The causes are the factors and are often called independent variables, and the effects are exhibited in terms of properties or attributes that vary depending on the factors; therefore, they are often referred to as dependent variables.

3.2.1   Variable types

Researchers should be clear about which attributes or properties are important to collect or generate for comparisons or to look at associations. Addition of a single variable may add considerable amounts of cost, effort, and time spent for the research. Some variables are qualitative and easy to collect, e.g. race, gender, occupation, and are called nominal variables. Many others are quantitative and called measurement variables as these can be measured and expressed in a numerically ordered fashion. The quantitative variables are also known as ratio or interval variables as they can be fitted or compared on a numerical scale. In some cases, there are attributes that can be compared but may not be easy to quantify. An example is comparing rich and poor farmers; this proves difficult unless all of their assets are converted into monetary terms, which is not easy to do, and in real situations, people may not bother to do so. How much richer the group is compared with the poor group may not be clear. These types of variables are often classified as ordinal variables. Even though they are not quantifiable, as statistics deals only with numerical facts, all qualitative information has to be converted into numerical form before analysis, using ranks or assigned numbers for the group differing in particular characteristics.

A variable after enumeration is also called data or numerical fact, which can be a continuous or discrete series. The continuous series has infinite intermediate levels in between two points, whereas a discrete or meristic series has none. For example, weights of fish can be in between 1 kg and 2 kg, e.g. 1.3 kg, 1.8 kg, or even 1.35 kg or 1.82 kg, whereas if 1 and 2 are the number of fish counted, there cannot be 1.3 or 1.8 fish. There are either 1 or 2 fish in a bucket. In the first case, data are in a continuous series, whereas in the second case, the data are in a discrete series.

Enumeration of qualitative information is quite difficult in many cases, therefore it should be done carefully. These attributes or nominal categorical variables are arbitrarily given numbers or ranks to present the group and make analysis possible using statistical tools. For instance, if 1 represents red and 5 represents white, then 2, 3, and 4 can be given depending on the degree of color in between these two. Similarly, numbers can be assigned for groups, such as 1 for the best taste, 2 for good, and 3 for fair, to be given by the taste panelists when comparing the tastes of fish from among the strains, source, or methods of cooking.

3.2.2   Variables of aquaculture

There are a number of variables used in aquaculture research. For an aquaculture trial, the main measurable variables are the number and weight of fish, fingerlings, or eggs (individual or batch) based on which other parameters are computed and used for comparisons. The computed parameters using two or more measurable variables are often called derived variables. An example of a derived variable in aquaculture is survival rate, or the number of animals that survived in relation to the number originally stocked. This is one of the most important derived variables in aquaculture and other animal production systems.

Other important derived variables are growth and productivity. For example, daily weight gain (DWG, g·fish·day^{-1}) is the rate of growth assumed to be linear throughout the research period, which is more or less true in grow-out phase (Phase II in Figure 3.2).

$$DWG = (W_1 - W_0)/T$$

Where,

W_1 is final mean weight,

W_0 is initial mean weight, and

T is time period or number of experimental days.

At younger stages, most organisms grow exponentially. Therefore, the appropriate variable or parameter for growth is to report specific growth rate (SGR) as a percentage, which can be computed using natural log (L_n) as:

$$SGR(\%) = 100(L_n W_1 - L_n W_0)/T$$

On the other hand, to compare the growth of broodfish, usually relative weight gain (or loss) is used, which means the percent gain or loss in weight is measured against the original weight. It is important to understand that this gain or loss may not occur only on a daily basis. For example, a female fish may lose a lot of weight just after spawning, which means it may occur within 1 minute. Most organisms normally have three distinct growth phases, which is explained by

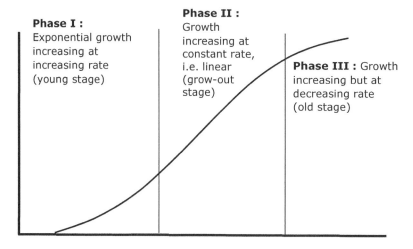

Figure 3.2 Growth pattern in aquatic organisms.

asymptotic function (Figure 3.2). Therefore, it is important to have as many intermediate data points as possible so that the actual growth curve can be fitted and comparisons can be made not only at the end point. Similarly, productivity of a system or yield is usually reported in terms of net fish yield, which measures the increased growth per unit area or volume per day. In general, ton·ha^{-1}·year^{-1} is used for larger culture systems (especially earth ponds), whereas kg·m^{-2}·day^{-1} or kg·m^{-3}·day^{-1} is used for smaller culture systems, such as tanks and cages.

Reproductive performance is expressed as the total number of eggs or young ones produced by each female. It is expressed as fecundity. Total fecundity is determined by estimating the number of eggs per female per spawning. However, it can be misleading, as the sizes of females and their ovaries differ considerably within the same group. Therefore, relative fecundity, i.e. number of eggs per kilogram of female per spawning, is used as a parameter for comparisons. However, reproductive performance of frequent spawners collectively managed (group as the experimental unit) has to be estimated per unit time period, i.e. day, week, or month, rather than per spawning; for example, in tilapia, the number of eggs, fry, or fingerlings (collectively called seed) per female or per kilogram of female per day or week. As tilapia broodstock are managed in groups and require considerable space, their reproductive performance is often measured per unit area of culture system, i.e. number of seed per square meter per day (Bhujel et al. 2007).

For nutritional research, feed conversion ratio, net protein utilization, and other feed efficiency parameters are used. It is quite common to see the effect of nutrients on the quantity or quality of reproductive cells, e.g. eggs/ova and sperms by counting or measuring their lengths and/or diameters. In addition, the rate of feed converted to egg biomass might be an important parameter if the purpose of trial is to analyze the reproductive performance. If these data are in percentage form, data have to be transformed. However, qualitative parameters, e.g. color and shape, are quite commonly observed but not analyzed. Similarly, taste of wild versus cultured fish can be compared. As these data are qualitative in nature, statistical analysis is possible only when they are assigned or coded with numbers. These types of data are analyzed by using nonparametric tests; some of the important ones have been covered in this book.

Many parameters are related to each other. For example, survival of fish has an effect on daily weight gain, and both of them have an effect on net fish yield. Therefore, in such a case, multivariate analysis is needed. However, most of the researchers analyze these parameters separately. In addition, most aquaculture researchers work hard to collect water-quality parameters, such as water temperature, pH, dissolved oxygen, nitrite, and ammonia levels, but these are merely analyzed. They might mask treatment effects in many cases. These could be used as covariates as they cannot be controlled, are also called noise factors, but have considerable effects on survival and many production parameters. If they are used as covariates during analysis, their effects can be separated. It could increase the chances of determining the real treatment effects.

Normally, economic parameters are not used to compare statistically among treatments. As it is quite difficult to keep records separately for each replicate, it is considered one of the most difficult parts of the analysis. Economic parameters are important only when there is a significant difference between treatments. Normally, economic figures are determined for the treatment combined (collectively for all the replicates of the treatment) and compared between the treatments rather than computing separately for each replicate, computing variance for comparison in a statistical way.

3.3   Fish sampling

A researcher can weigh the whole group (batch weight per replicate) and divide by that number to estimate the mean weight, or sample a certain number of fish to weigh individually and compute the sample average as a representative weight of the whole replicate group. The first method gives a more accurate mean; however, in many cases, it is not possible to weigh all of the fish. In such cases, sampling of a certain number of fish is necessary. Live fish sampling is quite difficult to do, and it is difficult to determine whether a proper random sample can be drawn because fish move fast from one corner to another. If the experimental units, e.g. ponds or tanks, are quite big, although it is difficult, it should be kept in mind that samples need to be taken from the middle parts as well as from every corner. Most researchers do carry out the intermediate samplings but do not use data of intermediate samplings and use only final or end point data for statistical analysis. If there are intermediate samplings, the data collected can be valuable and informative to see the trend over the period rather than just the result at the end. In such a case, time should be considered as block while performing statistical analysis. Therefore, separating the block effect would increase the reliability of the test or decrease the chance of committing an error.

3.4   Sampling of feed and feed ingredients

Sampling of feeds and feed ingredients is quite difficult, especially if it involves large volume. Sampling from commercial feeds might be easier because every pellet or mass is supposed to be uniform. But, if sampling is taken from homemade feeds, extra care should be taken in order to make the samples representative. If samples are to be taken from large volumes, multistage sampling (see Section 3.1) should be applied. As the conditions of storage have effects on the nutrient loss, sampling should be considered and representative subsamples should be taken from each part, e.g. bottom of the stock, middle and top of the piles, and similarly, from dark corners and well-lit areas. If the storage loss of different types of feeds or ingredients has to be compared in such a case, samples are kept separate and the factors can be considered as factor (block) while analyzing the data.

Feed samples should be dried and stored in deep freezers if they are analyzed after a few days or even later. While preparing for analysis, feed samples have

Figure 3.3 Feed samples in a feed company's lab in Mekong delta, Vietnam.

to be well-ground and mixed uniformly. Figure 3.3 shows a large number of samples ready for analysis in a feed company's lab.

3.5   Water sampling and monitoring

The environmental factors to be considered are water temperature, dissolved oxygen (DO), pH, turbidity, chlorophyll-*a*, ammonia, nitrate, and nitrite concentrations. Within a pond, these parameters fluctuate with diel cycle. Monitoring of water quality, early in the morning between 06:00 and 06:30 hrs for the lowest and in the afternoon between 14:00 and 16:00 hrs to record the highest levels, is important, especially for temperature and DO over the experimental period. Most water-quality parameters usually vary with depth; therefore, sampling or measurement should also be taken at the depths of 20 and 50 cm. But if the purpose is to see the variation with depths, then measurement could be done at 10-cm intervals. For representative sampling, a simple water sampler (Figure 3.4a) can be locally made using PVC pipe, the bottom of which can be covered once the sampler is brought down slowly into the water to collect water from the whole column. Usually, 100–200 mL of water, which can be stored in small, plastic bottles (Figure 3.4b), is sufficient for an analysis, but if there are other parameters to analyze, then 1–2 liters of water need to be sampled from various points. However, many researchers get puzzled about the volume of water to be

(a)

(b)

Figure 3.4 (a) Locally made column water sampler. (b) Water samples collected in plastic bottles stored in a freezer.

sampled. They may think that only 1–2 liters of sample might not be representative, especially when trials are conducted in ponds that contain large volumes of water. Fortunately, water is more uniform as compared with bottom soil, which differs even within a centimeter of distance.

3.6   Sampling of eggs, muscles, blood, and others

Depending on the purpose of the analysis, samples of eggs (Figure 3.5), muscles, blood, and other parts of fish have to be stored in a deep freezer (at −20°C) immediately after collection. Care must be taken so that these samples represent the population. For lipid analysis, for example, the sample should be preserved by using chloroform:ethanol (2:1) and butylated hydroxytoluene to protect from

Figure 3.5 Samples of different stages of tilapia eggs collected for laboratory analysis.

oxidation of fatty acids. Similar standard protocols need to be used for other purposes. If the samples have to be transported, they must be kept in a sealed icebox. In cases of transporting samples across country borders, a researcher may need to have specified documentation as they are considered to be live animal cells.

Samples of eggs, muscles, blood, and other parts are used for the purpose of comparing changes in chemical or nutritional compositions, especially for lab analysis. Their compositions might be affected by the same treatment factor, which means that they might have correlation. At the same time, regression analysis with the treatment factor and correlation among the dependent factors indicate the right method of analysis. In such a case, multivariate analysis is the correct tool, rather than analyzing the single parameter individually.

3.7   Sample size (volume/number)

Deciding the size of samples, both volume and number, is one of the most important, and also somewhat difficult task. Sample size has a direct impact on the amount of effort and the costs of research. Many researchers decide on an ad hoc basis, which results in too much work in some cases and shortage of samples at the end in other cases. Therefore, prior determination of the actual volume necessary for analysis is suggested. It is also strongly suggested that an additional volume of the sample, e.g. at least double the amount or even more, is taken and stored in the proper place and condition, so that if there is any failure of equipment, electricity, or anything else while analyzing the samples, stock can be used to repeat it. This is particularly important in cases where there is a limited time or window of opportunity for sampling.

In cases of sampling of water, soil, muscle, feed, and feed ingredients for analytical purpose, usually a small volume is adequate. For example, 5–10 g of each feed or its ingredients is enough for proximate analysis. Similarly, 100 mL of water is more than enough for most of the laboratory analysis. However, a researcher should be certain that these volumes really represent the whole from where they are taken. In these cases, multistage sampling would be the best method. The basic principle is to collect the largest volume possible at the first stage so that it represents the whole volume of water of a pond, tank, or other experimental unit. At the same time, larger volume will be less affected in case there is any contamination while handling the samples during analysis or sample preparation.

In most cases, sample size may mean the number of animals or objects that represent the whole population of any experimental unit. Most researchers often decide based on the percentage, e.g. 20%, 10%, 5%, or 3%. But when there are situations in which researchers need to make decisions about sample size from sufficiently large populations, e.g. tens of thousands, hundreds of thousands, or even millions, then a percentage formula would not work because even 0.1% would result in a large number that is not manageable. Two methods have been found to determine sample size.

3.7.1 Simple method for sample size estimation

In this method, as mentioned by Knud-Hansen (1997), sample size does not depend on the size of the population, but instead depends on the potential variance among the individuals in the population and the method of sampling, which determines how far the sample mean can be expected to be the population mean; in other words, an acceptable difference between sample mean ($\bar{X}$) and population mean (m). Using the following equation for *t*-test, sample size (*n*) can be estimated as:

$$t - \text{statistic } (t_\infty) = (\bar{X} - \text{m})/s \sqrt{n}$$
$$\text{Therefore,} \quad n = [(t_\infty \times s)/(\bar{X} - \text{m})]^2$$

For example, if a researcher found a standard deviation (*s*) of 16 g from a preliminary sample (or from published literature with similar type of research) from

a tilapia trial and the difference between population and sample means ($\bar{X} - m$) of 5 g, then the sample size can be estimated as:
Here,

t_∞ = value from *t*-table assuming ∞ *df* at $0.05 = 1.96$
$s = 16$ g
$\bar{X} - m = 5$ g

Therefore, the minimum size of the sample $(n) = [(1.96 \times 16)/5]^2 = 39.3$, or rounded to 40.

Therefore, the minimum sample size should be 40. It is strongly suggested to take a larger sample than this to ensure that at least 40 individuals will remain at the end, in case some of them die or become unusable due to unavoidable circumstance, such as handling and transportation. For example, if 10% mortality during handling is common, then:
Actual size of sample $(n) = 40 + (10\% \times 40) = 44$ per experimental unit.

3.7.2   Comprehensive method for sample size estimation

A more comprehensive method of computing sample size is using power of the test. As in the first method, standard deviation is obtained by presampling or is taken from related studies, and minimum detectable difference is assumed. In this method, probabilities of committing both errors (Type I and Type II; for more information, see Chapter 6) are taken into account. If b is the probability of committing Type II error, then statistical power is $1 - b$, which is the probability of detecting the significant difference or correctly rejecting a false null hypothesis. The following equations described by Zar (1996) can be used to estimate the sample size for one sample *t*-test assuming a minimum acceptable level of statistical power, i.e. 0.80, and at the same time the power of the statistical test can be computed back using the sample size if it is known.

$$\text{Sample size } (n) = s^2/d^2 \quad \times (t_{a,df} + t_{b,df})^2$$

Where,

n is number of samples or replicates,
s is standard deviation of sample from presampling or similar past studies,
d is minimum detectable or meaningful difference,
df is degree of freedom,
$t_{a,df}$ is significance level (e.g. 0.05), and
$t_{b,df}$ is power of the statistical test (e.g. 90%).

Similarly, this equation can be used to determine the power of the statistical test; for example, if $d = 1.0$ g, $n = 12$, and $s^2 = 1.5682$ g^2, then the power of the test can be computed as:

$$n = s^2/d^2 \times (t_{a,df} + t_{b,df})^2$$
$$t_{b,df} = d \div \sqrt{(s^2/n)} - t_{a,df}$$
$$= 1 \div \sqrt{(1.5682/12)} - 2.201 = 0.57$$

From the *t*-table, 0.57 corresponds to about 0.25, which is the b; therefore, power = 1 − b = 1 − 0.25 = 0.75, which means there is a 75% chance of detecting significant difference, which is lower than the normally acceptable level (i.e. 80%). Low statistical power means that sample size/replication is lower than the required, which might result in effects of a treatment going without detection/ notice.

Similarly, for two-sample *t*-test and analysis of variance (ANOVA), sample sizes and the statistical powers can be computed using the following equations:
For two-sample *t*-test: $n = 2s_p^2/d^2 \times (t_{a.df} + t_{b.df})^2$
Sample size and power of ANOVA:

$$\Phi = \sqrt{\frac{nd^2}{2ks^2}}$$

Where,
   Φ is statistics based on which to see probability from *F* table,
   k is the number of treatments/factors,
   d is minimum detectable difference, and
   s^2 is variance.

3.7.3   Sample size estimation for survey research

This method has been suggested by Yamane (1967) for research based on the social survey. In this method, sample size depends on the size of the population; however, sample size does not proportionately increase with the increase in population size. Even if the population is very large, the sample size does not go beyond 400; however, probability of committing Type II error (b), another factor affecting the sample size, should be considered lower, i.e. 5% (90% confidence level), than the normally used probability, i.e. 10% in the equation.

The equation for sample size $(n) = N/(1 + N \times e^2)$
Where,
   n is sample size,
   N is total population, and
   e is probability of committing Type II error or b (normally 10%).

For an example, if there are 400 fishers' families in a village, then for the number of sample households required $(n) = 400/(1 + 400 \times 0.10^2) = 80$.

In many cases, at least for some parameters, some of the families might not respond or data can be missing. In order to compensate for that, 5–10% more should be considered for survey. Therefore, the actual sample size $(n) = 80 + (80 \times 10\%) = 88$.

Many researchers get confused with the sample size and the number of households to be surveyed or interviewed. Before making a decision, they should be clear about the research objectives, which determine the number of samples and households needed. For example, if the objective of a research is to compare the parameters of a particular village against the national standard parameters established by the government or any organizations (statistical bureau or alike) using standard methods, then there is only one sample involved in the research; therefore, the number of households to be surveyed is the sample size of that village as estimated above. But, if the researcher is to compare parameters between two villages then he or she should estimate the sample size for each village separately. That means the total number of households to be interviewed is the sum of the sample sizes, which can be double. Similarly, if the purpose of a research is to compare a parameter between two ethnic groups of fisher folks living in the same village, then a reliable source of information should be obtained or a preliminary survey is necessary to know the approximately numbers of households in each ethnic group so that actual sample size can be worked out for each group. In reality, it is very unlikely that two ethnic groups in a village or two separate villages can have the same population. Therefore, the size of the samples would be different. But, most researchers try to set the number equal, either because they think it will be easier for data-handling purposes or they lack statistical knowledge. If anyone chooses to use an equal sample size, then he or she has to pick the highest figure of the villages under study.

3.8   Questions

Q1. Why is sampling important?

Q2. You are supposed to recommend either tilapia or catfish for culture. In order to compare between two species, which variables would you use for comparison?

Q3. If you are a quality control officer responsible for a district where there are 30 feed manufacturers producing four types of feeds, how would you perform sampling for laboratory analysis of the feeds they produce?

3.9   Practical exercises

Ex. 1. Calculate the total number of fish to be sampled from 20 experimental tanks (5 treatments with 4 replications) containing 1,000 fish each, if you expect the difference of 10 g.

Ex. 2. Determine the total number of fish farmers needed to be interviewed for a research to compare productivity of three cooperatives (A, B, and C) that have 350, 1,000, and 1,500 members, respectively.

Ex. 3. Data shown in Table 3.1 are the batch weights (g) of 15 fish from each replicate tank. The experiment was conducted at a fish hatchery using circular tanks (1.75 m^2) for 15 weeks. Compute the following derived variables and their treatment means for each treatment using the formulas given below. Present them in tabular forms and bar diagrams.

a. Daily weight gain (weight·fish^{-1} · day^{-1})

b. Relative weight gain = (Final weight – Initial weight) × 100/Initial weight

c. Specific growth rate = $(L_n W_1 - L_n W_0)$ × 100/no. of experimental days

Table 3.1 Batch weights of 15 fish from a trial at AIT, Thailand.

Treatments	Feeding Rrate (%)	Replication	Batch Weight (g)	
			Stocking	Final Weight
A: Normal fish	1	1	663.3	784.2
(3- months old)	1	2	595.7	840.6
	1	3	581.6	814.9
	2	1	548.4	1005.1
	2	2	636.3	1304.8
	2	3	643.2	1259.1
	3	1	609.0	1513.9
	3	2	636.1	1432.4
	3	3	661.6	1291.3
B: Stunted fish	1	1	588.7	790.6
(12- months old)	1	2	493.5	734.8
	1	3	549.0	750.6
	2	1	505.5	1179.5
	2	2	517.8	1007.2
	2	3	549.3	1151.7
	3	1	560.2	1612.9
	3	2	526.6	1370.2
	3	3	572.8	1608.7

Note: normal fish (Group A) were only 3 months of age, whereas stunted fish (Group B) were 12 months old but raised at high density under limited, restricted feeding conditions.

d. Net fish yield $(g \cdot m^{-2} \cdot day^{-1})$ = (Final biomass − Initial biomass)/area of tank/no. of days

e. Net fish yield $(t \cdot ha^{-1} \cdot year^{-1})$

Chapter 4

Data accuracy and exploratory analysis

4.1 Importance

Exploratory analysis implies finding any indications or trends and also pointing out any errors in data sets. In most of cases, collection of data from the field or laboratory is carried out by technicians who may not be aware of the treatments and their effects or the objectives of the research. They just do their regular job and follow the set of guidelines. They may not be careful about the accuracy of data and precision because they just record whatever they measure. However, in some cases, data might be recorded by those who are very much aware of the research objective, hypothesis, or the treatments, e.g. student research. When they record data, they are very careful; therefore, they can have some prejudice toward supporting their hypothesis. In the former case, there is a greater chance of unbiased data, but there can be more errors. Whereas in the second case, errors are avoided but there can be biased data. All data should be carefully checked before keying. Any odd data should be carefully handled. If there are problems, the causes should be found so that any outliers (extreme points) can be either corrected or rejected with adequate justification.

Exploratory data analysis is very important as a single incorrect datum may result in skewed means and medians hence need of repetition of the whole analysis. Therefore, actual statistical analysis should begin only when a researcher is fully confident that there are no errors or odd values.

4.2 Data accuracy and precision

Accuracy implies the nearness of a measurement to the actual or expected value of the variable. Appropriate units and their levels of measurement (kg, g, mg, mg, etc.) for each variable have to be selected so that the data recorded can be as accurate as possible. For example, the weight of a fish could be measured

Neither precise nor accurate   *Precise but not accurate*   *More accurate and precise*

Figure 4.1 Diagram showing distinction between accuracy and precision.

up to one decimal figure, e.g. 8.5 g, or up to two decimals, e.g. 8.53 g, and so on, depending on the level of accuracy of the measurement needed and also the capacity or type of measuring instrument. Precision has been misunderstood quite often. It means closeness of repeated measurements or data points to each other. But, if any factor is affected during measurement, data points can be close but they may not be accurate. Therefore, we strive for both accuracy and precision. Figure 4.1 explains the distinction between accuracy and precision.

4.3   Significant numbers

Although it sounds very basic, the author feels the need to mention that most researchers are not careful or clear about significant numbers when recording data and reporting results. The level of accuracy in recording the data determines the type of equipment/instrument required. For example, to weigh fish fry, a balance that can measure up to two decimal points of a gram is necessary, whereas to weigh bigger fish (100 g or above) for grow-out trials, measuring up to two decimal points is not necessary. In such a case, simple balance measuring between 1 g and 1000 g would be enough.

The basic principle of precise measurement is that there must be enough space to exhibit variations in data so that statistics can detect them and determine whether the differences are significant or not. In general, between minimum and maximum values expected to measure, there should be 30 to 300 intermediate levels. For example, if we expect that weights of fish in our trial will range from 5 g to 10 g, then a balance that records grams without decimal figures should not be used as there will only be 5 intermediate steps between 5 g and 10 g. We need a balance that can record up to one decimal figure, which means recorded data will resemble 5.0 g, 5.1 g, 5.2 g, etc., up to 10.0 g. There will be about 51 steps between the minimum (5.0) and maximum (10.0) values. A balance that can measure up to two decimal points is also not necessary. There will be 501 intermediate steps if measured between 5.00 g and 10.00 g. Therefore, measurement or data recorded up to one decimal place is adequate in this case. The same principle should be applied for other units as well. If data have been recorded more precisely than needed using decimal levels, then they need to be rounded off to keep only the significant figures. It is a general rule that calculated or derived variables can have one more digit. For example, the average family size of a district can be 3.5 even though number of family members cannot be 3.5 in reality.

Similarly, when computing figures, answers should not be expressed more accurately than the least accurate figure used. For example:

1. $5,200 + 85.7 = 5,285.7$ (incorrect) $=> 5,300$ (correct)
2. $5,200.0 + 85.7 = 5,285.7$ (correct)
3. $5.15 \times 3.1216 \times 150 \times 561.617 = 1,354,303.452$ (incorrect) $=> 1,354,300$ (correct)

Table 4.1 shows examples of calculated values that can be reported a variety of ways, depending on how many digits are significant. It is suggested to write out numbers in words if there are many zeros after the rounded numbers which are not significant, e.g. 5.6 million would be better or shorter than writing 5,600,000 or 5 million and 6 hundred thousand.

Table 4.1 Examples of significant figures.

SN	Calculated Values	Four Significant Digits	Three Significant Digits	Two Significant Digits	One Significant Digit
1	313.62	313.6	314	310	300
2	5,572,841	5,573,000 (5.573 million)	5,570,000 (5.57 million)	5,600,000 (5.6 million)	6,000,000 (6 million)

4.4　Errors and their sources

Errors in data can mask the treatment effects, easily leading to faulty conclusions and recommendations. Therefore, it is extremely important to take measures to avoid, minimize, or separate errors, so it is important to know about their sources. Three main types of errors are described in the following sections.

4.4.1　Gross errors

These errors are due to incomplete or missing data, missing important persons/times (e.g. DO should be monitored at around 6 a.m. but is likely to be missed), malfunction of the instruments while recording data, human errors, intentional data manipulation, mistakes while typing/keying, contaminated reagents, and so on. Due to these errors, data become neither accurate nor precise; therefore, these errors should be avoided or minimized as much as possible.

4.4.2   Systematic errors

The errors that occur repeatedly due to bias, rounding off, and faulty calibration of reagents and instruments are called systematic errors. In the presence of these errors, data can be precise, but they are not accurate. It is possible to separate, avoid, or minimize these errors by adjusting, revising, or recalculating the data recorded with errors, with proper evidence of errors.

4.4.3   Random or residual errors (unsystematic)

These are the remaining errors which vary unpredictably. It is impossible to completely wipe out all errors as every individual/object differs from one another. For example, even identical twins differ in physical appearance and mental capabilities. The errors are often called experimental errors, which are the basis for comparison. Effects of treatments or blocks have to be sufficiently higher than the random errors to be significant.

4.5   Error minimization and separation

Gross and systematic errors can be avoided if the researchers plan properly, use proper sampling methods, keep control of the trial or research project, and avoid re-keying of data. Many researchers enter data again or take rounded figures when they need to calculate and compare another derived variable. In such cases, the chance of making errors increases. Therefore, it is strongly advised to copy and paste from the original data file if they have already been entered once. Residual error is impossible to avoid, but it can be minimized to the smallest error possible. The following seven rules would help reduce errors.

4.5.1   Experimental conditions and procedures

When/wherever possible, it is wise to preset and prerun the trial so that data generated during the trial will have less variability. Pretesting of instruments, equipment (e.g. DO and pH meters), reagents, or any other chemicals, experimental systems, or questionnaires is necessary.

4.5.2   Materials, methods, and equipment

It is necessary to use the same size and age of fish or experimental materials (units) as much as possible for the trial if the size is not the treatment. Similarly, the same method should be used for the whole trial period, e.g. using DO meter to measure DO level in 1 week (treatment) and chemical method in another week means there is a high chance of variability. When/wherever possible, the same instrument or at least the same model should be used throughout the experimental period.

4.5.3   Randomization

This is the cornerstone of the statistical theory in the design of experiments as it provides similar and equal chances to all of the treatments. Randomization is normally done using lottery method or random numbers by using a table or generating spreadsheet programs, e.g. MS Excel [=Rand ()*1000]. A random table of any size can be generated whenever needed. An example of a 10×20 table generated is shown in Table 4.2.

Suppose four treatments, T1, T2, T3, and T4, are to be randomly allocated in 8 ponds that are in 2 rows. A number can be chosen as the beginning point without looking at the random table. Let's say 368 (see Table 4.2) was chosen by chance; then take the other three figures, i.e. 271, 499, and 799, on the same row, and take the four figures right below them, i.e. 409, 901, 76, and 920. These two rows can be considered as two blocks. The first-row tanks are numbered 1–4, and the second-row tanks are numbered 5–8, then T1 goes to the tanks with the lowest random number in each row, which means tanks with the number 271 in the first row and 76 in the second row. Similarly, T2 goes to the second lowest random numbers, which means tanks with 368 in the first row and 409 in the second row. Similarly, T3 and T4 can also be randomized. If a researcher feels that

Table 4.2 A sample of random table generated using Microsoft® Excel spreadsheet using function = random*(1000).

	1	2	3	4	5	6	7	8	9	10
1	809	377	965	597	417	778	475	61	692	341
2	153	33	466	147	86	226	301	636	475	700
3	521	361	824	347	165	140	978	480	755	16
4	174	240	987	24	27	151	109	686	815	702
5	992	879	366	980	116	923	595	489	746	588
6	133	752	967	273	259	263	609	285	547	977
7	312	209	983	788	992	202	266	835	227	952
8	552	454	16	138	665	**368**	**271**	**499**	**799**	614
9	182	155	469	288	20	**409**	**901**	**76**	**920**	336
10	417	203	324	888	624	431	61	796	612	425
11	97	166	487	938	958	523	914	518	258	259
12	511	68	40	960	22	719	901	360	204	201
13	268	701	71	299	794	643	764	582	468	670
14	341	369	627	2	994	246	811	283	354	218
15	234	60	8	413	234	328	255	721	881	245
16	420	889	291	883	74	561	287	201	527	383
17	936	69	847	271	184	480	645	46	668	487
18	389	146	178	188	742	631	529	607	456	141
19	603	94	600	926	281	791	743	772	832	114
20	163	207	740	762	55	183	316	823	762	496

the initial randomization does not look well-randomized, then it can be repeated to get the best randomized layout. More specific details on randomization for each type of experimental design are described in Chapter 7.

4.5.4   Pairing

Pairing means grouping of experimental units into two. It is extremely difficult to find same-size animals, even from the same age group. In such cases, the effect of size difference can be separated by pairing them based on size and assigning both treatments. Other examples include monitoring of water-quality parameters at the same time, such as DO, temperature, and others over time. The data obtained from paired experimental units are analyzed using paired sample t-test (for details on statistical analysis, see Section 6.8.1.4).

4.5.5   Blocking

Blocking means grouping of similar experimental units into a single group. Its purpose is to separate effects that already exist in the system which are either impossible or too expensive to avoid. These types of effects caused by factors such as canal, shade, different ponds/plots, districts, community, and so on can mask the treatment effects; therefore, they need to be separated while analyzing the data. A block can be spatial in area, as well as in terms of time.

4.5.6   Measurement of additional variables/factors

In reality, a biological system is very complex where several factors are acting together and separation of effects of the factors other than the treatment(s) is almost impossible. Any changes expected in certain experimental conditions that may affect the variables in question should be measured. Their effects can be separated by using covariance analysis. For example, water temperature, pH, and DO of aquaculture systems may vary with time, which affects fish growth and survival. Therefore, it is necessary to measure these types of variables even when they are not considered as treatments.

4.5.7   Increasing number of treatment and replication

Increasing the number of replication and/or treatment levels in experimental designs for regression or factorial analysis could improve precision and accuracy. Experimental error can be measured only if there are at least two units treated the same way. If we see the same thing happening again and again, we are more confident that that event happens again if such conditions are available. The minimum replication is two, but using only two replications has a risk and is considered unacceptable. Basically, the higher the replication, the more reliable

Table 4.3 Effect of replication on variation

	Trial 1	Trial 2	Trial 3
Treatment A	5 rep	6 rep	8 rep
Treatment B	5 rep	4 rep	2 rep
Standard Error (SE)	$SD\sqrt{(1/5 + 1/5)}$ $= SD \times 0.63$	$SD\sqrt{(1/6 + 1/4)}$ $= SD \times 0.65$	$SD\sqrt{(1/8 + 1/2)}$ $= SD \times 0.79$

the research outcomes will be. However, trials with higher replication are expensive, and sometimes facilities are not available. A range of four to eight replicates has been suggested for agricultural research; however, most aquacultural research uses three replicates. The basic principle is that, if very little variation is expected, then only three replicates will be sufficient and acceptable, but if a researcher suspects that there will be less control over other factors and variation will be high, then more replication should be used. Care should be taken with treatments, replications, and experimental units. For example, use of nitrogenous fertilizer in a fish pond is a treatment (control: without the use of nitrogenous fertilizer). Doses of nitrogen, e.g. 20, 30, or 40 kg·ha^{-1}·week^{-1}, are treatment levels. Researchers should be careful about pseudoreplication. Replicate samples or subsamples measuring individual fish in a tank is pseudoreplication if the tank is an experimental unit. In a grow-out trial, tanks, cages, or even ponds can be experimental units. On the other hand, in a breeding trial, individual fish can be experimental unit as the variables will be recorded based on the individual performance. While in a grow-out trial, all the fish in a tank, cage, or pond are considered a group, and average weight of the whole group is used to compare with the average of another treatment group. Treatments can either be replicated spatially or in terms of time. If the experimental units are not sufficient, they can be replicated over time, e.g. weekly, monthly, yearly, which may provide additional information on temporal or seasonal variations.

Replication can be different for different treatments, but equal replication decreases the standard error or variance and increases the precision (Table 4.3).

A minimum number of replications or samples can be calculated if we know the expected variance and minimum substantial difference from a preliminary sampling or similar research carried out in the past using the method described in Section 3.7.

In field trial, there is a possibility of not having enough replication, which may give nonsignificant results; therefore, power analysis is needed to determine whether the nonsignificant difference was due to inadequate replication or the real treatment effects.

4.6   Exploratory data analysis

There are several steps and methods for exploring initial trends, outliers, and others before analyzing any data. Exploratory data analysis should actually begin from the beginning when data are starting to be recorded. An experienced person can immediately point out when there is any extreme value, discover the possible reason(s), and write some notes on the data sheet. Therefore, while managing the experiment and data recording systems or equipments, data sheets should be separate and specific to each variable, e.g. a sheet for a day which can be filed chronologically. The data sheet should include records of the person (who collects the data), date, time, and any other conditions, so that all data can be traced back in case of doubt during exploratory data analysis, or even any time in the later stages.

The following major steps are suggested before starting proper statistical analysis.

4.6.1 Checking for any errors

- obvious mistakes: double check original data, ask someone else to check (you may not see your mistakes)
- precision of recording
- recorder/instrument differences
- trends with treatment levels and time, any increase/decrease
- treatment responses
- extreme values

4.6.2 Comparison with others

Comparing data helps researchers build confidence in their work if they find similar and relevant data in published forms, such as:

- Journals and magazines
- Books and proceedings
- Newspapers, thesis, reports, and other forms

4.6.3 Useful tools

- **Tables:** Drawing a table of a summary of results is a must, as it can accommodate a large amount of information to see at a glance. It can show the exact numbers, e.g. frequency and its distribution, cumulative frequency, sum, mean, maximum, minimum, etc., based on which results can be predicted.
- **Graphs:** The first step in exploratory data analysis for each variable starts with drawing scatter plots to see the distribution of data. Line graphs can be used to see/show the trends, whereas bar charts are for discrete series. If the data are in circular fashion, pie charts should be used. Depending on the nature of data and the objectives, various other types of graphs, e.g. frequency distribution polygons or histograms, can be used. Reports or papers should not be loaded

with many graphs. During the exploratory stage, graphs can be made for all of the variables, but they should be used only for very important findings in order to place more emphasis to them, especially in the final presentation or publication.

- **Pictures and diagrams:** A single good picture can describe something better than thousand words. Therefore, the use of pictures should be maximized wherever possible. But care must be taken, because space occupied by pictures and diagrams can be huge; therefore, their inclusion in presentations/publications can be problematic.
- **Unexpected events/data:** It is common that extreme values may be recorded or observed. Do not discard them, even if they are unexpected. Try to find the causes and solutions. Simple notes and explanations can be very important sometimes. For example, a member of the AIT outreach staff found a record of a fish pond 5 m deep. He was not sure about that. He went back to the field to check and asked the farmers whether that was correct. He found that it was correct because the farmer used the pit he had made when he used the soil to make bricks. A note from the data collector could have made this clear and saved the cost of a trip (time and efforts) incurred later.

4.6.4   Basic assumptions

Before starting detailed statistical analysis, it is necessary to determine whether the collected data from samples show any additive effects of treatments or whether blocks and errors are additive. For example, Table 4.4 shows a hypothetical example where use of nitrogen showed 33% and the block showed 50% additive effects.

Data or the observations collected from each replicate and their variances are supposed to be normally distributed or homogenous. Properties of normal distribution are described in Section 4.7. For tests for normality and heterogeneity of any data series, see Section 6.7 on the x^2-test and K-S test.

Table 4.4 Additive effect of treatments.

Treatment N/ha/wk	Block A	Block B	Block Effects	
			B – A	(B-A) × 100/A
T1: 40 kg	3.0	4.5	1.5	50%
T2: 80 kg	4.0	6.0	2.0	50%
T2 – T1	1.0	1.5		
T2 – T1/T1	33%	33%		
(Treatment effects)				

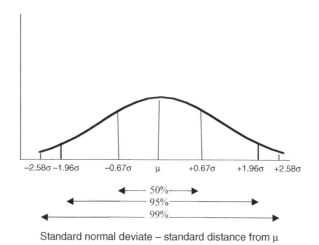

Standard normal deviate – standard distance from μ

Figure 4.2 Normal curve, area of coverage, and distances from the mean.

4.7   Normal distribution

If a set of data is normally distributed, the frequency distribution forms a graph, which is popularly known as the normal curve (Figure 4.2).

The normal curve is characteristically unimodal, bell-shaped, and symmetrical around the mean, i.e. skewness is zero and not too peaked or too flattened (kurtosis = 3). If the data series are perfectly normal mean, median and mode will be in the middle of the curve; or the data series are the same, and the mean is always higher than variance.

To know normality, researchers need to know non-normality. The following two words for non-normality are important to know:

- **Skewness:** the measure of asymmetry, i.e. pointedness of a curve toward the right or left. It is very hard to get zero (0) skewness, i.e. perfectly normal. Therefore, in general, data sets having skewness between +1 and −1 are considered within the normality. Negative value of skewness means the curve is pointed toward the left, whereas positive means toward the right. The top three curves in Figure 4.3 show skewness.
- **Kurtosis:** the measure of height or peak of a curve. In general, data sets with the kurtosis values between 2 and 4 are considered normal. If the height is higher than 4, it is called leptokurtic, and lower than 2 is platykurtic. The bottom two curves in Figure 4.3 represent these types of curves based on kurtosis as compared with the normally distributed one.

Table 4.5 is a summary which shows the normality ranges on the scales of skewness and kurtosis.

The following are the working formulas for skewness and kurtosis:

$$\text{Skewness } (g\,1) = (1/ns^3)\Sigma(X_i - m)^3$$
$$\text{Kurtosis } (g\,2) = (1/ns^4)\Sigma(X_i - m)^4$$

Where,
X_i is observations,
m is mean, and
s is standard deviation.

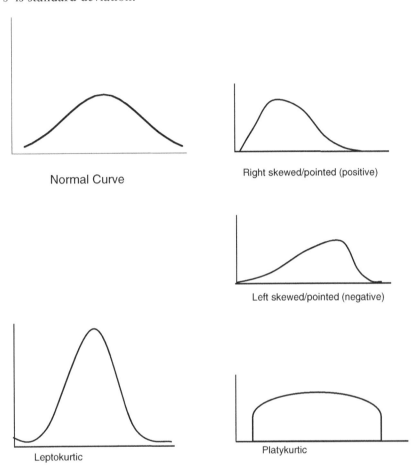

Figure 4.3 Types of curves based on skewness and kurtosis compared with the normal one.

Table 4.5 Summary of skewness and kurtosis in relation to normality.

Skewness (g_1)	< −1 Left skewed	0 Normal	> +1 Right skewed
Kurtosis (g_2)	< 2 Platy kurtic	3 Normal or mesokurtic	> 4 Leptokurtic

4.7.1 Concept of probability

In a population of fish, there is an equal chance of being sampled for male and female.

Here,

Probability for male (p) = probability of female (q) = 0.5

$$p + q = 1 \text{ (or 100\%)}$$

If a population has a mean (m) of 350 g and standard deviation (SD) of 15 g, what are the probabilities or chances of obtaining the following measurements?

a) 360 g or bigger
b) 380 g or bigger
c) 500 g and higher
d) Lower than 340 g and higher than 360 g?

Here,

a) 360 g

$$Z \text{ value} = (X_i - \text{m})/\text{s}$$
$$= (360 - 350)/15 = 0.67$$

Probability = 0.2514 = 25.14% (from table in Appendix A1)
b) 380 g

$$Z \text{ value} = (380 - 350)/15 = 2.00$$
$$\text{probability} = 0.0228, p = 2.28\%$$

c) 500 g

$$Z \text{ value} = (500 - 350)/15 = 10.00$$
$$\text{probability is} < 0.0001, p < 0.01\%$$

4.7.2   Frequency distribution and probability

Frequency is a count of repeated occurrence of a particular event or object, as shown in Table 4.6.

Discrete variable: Following are the data of the size of fish farmers' families collected by a researcher. Data have been arranged in a respective order to the family number, 1 to 20. Arrange them in a frequency table, draw a bar graph, and point out family sizes that have the lowest and highest number of families.
Here,
Family size (no. of family members): 5, 2, 3, 3, 4, 5, 3, 4, 4, 3, 4, 5, 2, 3, 2, 6, 4, 4, 6, 5

If a discrete variable also has large range (minimum – maximum), grouping is necessary. The number of groups to establish depends on the purpose and nature

Table 4.6 Frequency table showing frequencies of family size.

Family Size	Frequency (No. of Families)	Cumulative Frequency	Per cent (%)	Remarks
2	3	3	15	
3	5	8	25	
4	6	14	30	Highest
5	4	18	20	
6	2	20	10	Lowest

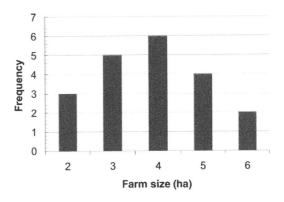

Figure 4.4 Frequency bar graph showing family size of 20 fish farmers.

of the data. However, the basic principle is that the number of group or class should not be too many or too few. The number of class should be between 8 and 12. For example, Figure 4.4 shows only 6 classes. Its bars look quite apart from each other. This shows that it could accommodate more bars, meaning additional information could have been included in the same space.

For example, the data set below is the number of tilapia recruits collected in each pond when harvesting of tilapia after growing for 6 months (pond no. 1, 2, 3, and so on): 25, 402, 203, 303, 204, 125, 38, 441, 200, 50, 112, 45, 200, 111, 0, 36, 14, 445, 60, 500, 1200, 300, 600, 20, 400, 30, 20, 22, 40, 300, 200, 1150, 300.

Find the range first and work out for class interval as:

Here,

Minimum value $= 0$

Maximum value $= 1200$

If class interval is 100,

No. of classes $= (1200 - 0)/100 = 12$

If we need to present these data in terms of pond size, such as small, medium, large, and very large, then we can use a class interval of 300 that gives only 4 class intervals.

Table 4.7 Frequency of discrete data series after grouping.

Farm Size (Class)	Frequency	Cumulative Frequency	Percent (%)	Remarks
>3	3	3	15	
3–4	5	8	25	
4–5	6	14	30	Highest
5–6	4	18	20	
>6	2	20	10	Lowest

Continuous variables: Unlike discrete variables, repetition of the same value is rare in this type of data set as there can be a number of intermediate values in between two. For the sake of simplicity, the values can be groups as shown in Table 4.7, and frequencies are shown for those groups.

Farm size (ha) (family no. 1–20, respectively):

5.4, 2.3, 3.5, 3.2, 4.5, 5.6, 3.2, 4.0, 4.4, 3.6, 4.3, 5.2, 2.3, 3.5, 2.5, 6.3, 4.5, 4.2, 6.2, 5.3

4.7.3 Grouping classes

An ungrouped distribution is a set of data that shows the actual values and observed frequencies for those values, whereas a grouped distribution is a set of data that shows frequencies for each group rather than the actual values recorded. Group is often called class. Each class has its extreme boundaries (not to overlap boundaries) called class limits. The number to the left of the class is the lower limit, whereas upper limit on the right. Some of the classes can be open, especially the beginning or the last class. For example, <25, 25–50, 50–75, 75–100, >100. The first class, <25, means any number even below 0 and the last class, >100, includes any possible values above 100, meaning values can even be 300 or higher. The difference between the true or mathematical upper and lower class limit (or difference in stated limits) is class interval, also called class width. There are two methods of grouping classes.

Method 1: upper limit excluded, e.g. the class 0–9 means 0 – under 9; that means data are included from –0.999 right up to 8.999, but 9 is not included.

Method 2: upper limit included, e.g. class 0–10 means it includes values from –0.5 upto 10.4. But not 10.5, which will be included in class 10–20.

4.7.4 Histogram and frequency curves

A histogram is a bar diagram for continuous data, a graph of frequency distribution with the x-axis extending from one class limit to the other and the observed frequency in the y-axis (Figure 4.5, left). The area of a rectangle is proportional to the observed frequency in the class. The vertical bars show the frequency

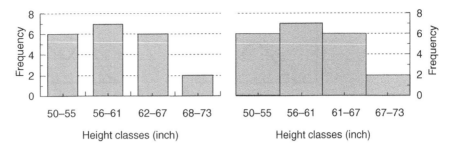

Figure 4.5 Frequency bar graph (left) and histogram (right).

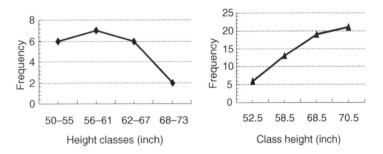

Figure 4.6 Frequency polygon (left) and cumulative frequencies (right).

density. When class limits are repeated, i.e. upper limit of the first class and the lower limit of the second frequency, the diagram looks like Figure 4.5 (right), popularly know as a histogram. A line graph drawn using mid-points (class mark) on the *x*-axis and frequencies on *y*-axis is called a frequency polygon. If the frequency polygon is smooth, it is also called a frequency curve (Figure 4.6, left). The shape of the frequency curve may change based on the class intervals. Using the frequency curves, it can be guessed whether a distribution is normal or not. A curve can also be drawn using cumulative frequencies (Figure 4.6, right) Cumulative frequency curve become straight if the data set is normally distributed. Therefore, based on the nature of cumulative frequency curves (Figure 4.7), it can also be guessed quite confidently whether they are in normal distribution or not. The maximum point of deviation is the reference point for K-S test of normality (see Section 6.8).

4.7.5   Variance heterogeneity

One of the characteristics of the normal distribution is homogenous variance. The nature of frequency curves (Figure 4.8) can also show whether variances are homogenous or heterogeneous. One of the indicators that has been widely accepted is the level of variance. If standard deviation (s) is greater than mean ($\bar{x}$), then the data set is considered to be not normal.

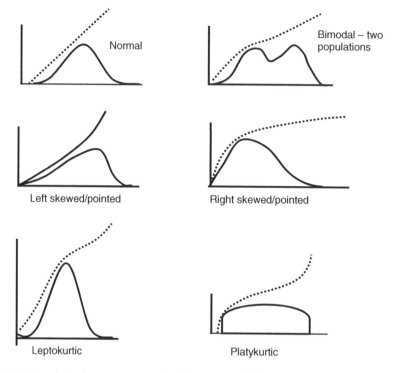

Figure 4.7 Cumulative frequency curves for different types of data sets.

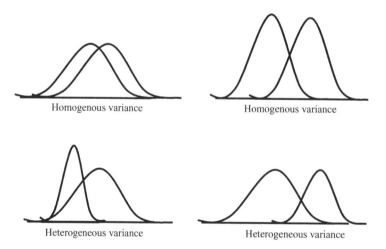

Figure 4.8 Frequency curves showing variance homogeneity/heterogeneity.

4.7.6   Data transformation

In most of the conventional statistical tests, such as Student's *t*-test, *F*-tests, or ANOVA, regression is designed for normal distribution. Due to the presence of a large number of uncontrolled factors, especially in the biological field, there are plenty of chances of having normally distributed data. If the data collected are not normally distributed, a researcher should either transform the data first to make them normal and analyze using parametric tests or they should be analyzed using nonparametric tests. Statisticians have also developed nonparametric tests. In this section, the main three methods of data transformation are briefly described:

- **Square root** ($\sqrt{X}$): Square root transformation is used mostly to count ratio or percentage data that have wide ranges. For example, survival rate of fry during nursing may range from 0% to 100% (e.g. Table 4.8). Percentage data between 30% and 70% normally do not need transformation. If variance is equal and proportional to the mean, square root transformation suits the best. In the case of percentage data with 0 values, 0.5 is added so that square roots can be obtained. Addition of 0.5 to all of the data is needed for analysis purpose only so it doesn't affect the results. Percentage data, therefore, will be in the form of $\sqrt{(X/100 + 0.5)}$.

- **Log or L_n**: Log or natural log (L_n, value 2.718282) transformation is suitable for the data with whole numbers having a wide range with multiplicative effects rather than additive. Log or L_n transformation changes this multiplicative effect to additive, which is the characteristic of normal distribution (see Table 4.9 for example). One of the characteristics of data when frequency curve is drawn is skewed to the right, and standard deviation is proportional to the mean or effects are multiplicative or exponential, which occurs in fast-growing organisms, e.g. during early stages of fish. Specific growth rate

Table 4.8 Square root transformation.

	Treatment A		Treatment B	
	Survival (%)	Sqrt	Survival (%)	Sqrt
1	50	7.07	15	3.87
2	40	6.32	25	5.00
3	55	7.42	25	5.00
4	60	7.75	30	5.48
5	70	8.37	40	6.32
6	80	8.94	30	5.48
7	90	9.49	20	4.47
8	100	10.00	25	5.00
Mean	68.1	8.169	26.3	5.078
Std	20.7	1.258	7.4	0.728
Var	428.1	1.583	55.4	0.530
Ratio of variance (V_A/V_B)			Raw data	7.7
			Transformed	3.0

Table 4.9 Log/L_n transformation.

	Treatment A				Treatment B		
	Survival (%)	Log	L_n		Survival (%)	Log	L_n
1	50	1.699	3.912		15	1.176	2.708
2	40	1.602	3.689		25	1.398	3.219
3	55	1.740	4.007		25	1.398	3.219
4	60	1.778	4.094		30	1.477	3.401
5	70	1.845	4.248		40	1.602	3.689
6	80	1.903	4.382		30	1.477	3.401
7	90	1.954	4.500		20	1.301	2.996
8	100	2.000	4.605		25	1.398	3.219
Mean	68.1	1.815	4.180		26.3	1.403	3.231
Std	20.7	0.135	0.311		7.4	0.127	0.292
Var	428.1	0.018	0.097		55.4	0.016	0.085
Ratio of variance (V_A/V_B)							
				Raw data	7.7	7.7	
				Transformed	1.1	1.1	

(SGR) $= 100\times (L_n W_1 - L_n W_0)/T$, which uses natural log transformation, is used for larval rearing or nursing trials, instead of DWG. If even an observation in a data series is less than 10, then 1 is added so that log transformation can be done, e.g. $= L_n/\log(X + 1)$.

• **ArcSine or angular transformation:** Angular (ArcSine), i.e. Asin(x), transformation has also been used to minimize the variation. It is normally coupled with square root transformation, i.e. Asin($\sqrt{X}$). In case of percentage data, if there are 0% values, they should be replaced by $1/4n$ and 100% by $100 - 1/4n$. Table 4.10 is given as an example.

Table 4.10 ArcSine-square root and then transformation to radians.

	Treatment A				Treatment B		
	Survival (%)	ArcSine	Radians		Survival (%)	ArcSine	Radians
1	50	0.650	37.2		15	0.398	22.8
2	40	0.591	33.9		25	0.524	30.0
3	55	0.675	38.7		25	0.524	30.0
4	60	0.699	40.1		30	0.580	33.2
5	70	0.742	42.5		40	0.685	39.2
6	80	0.780	44.7		30	0.580	33.2
7	90	0.813	46.6		20	0.464	26.6
8	100	0.841	48.2		25	0.524	30.0
Mean	68.1	0.724	41.5		26.3	0.535	30.6
Std	20.7	0.086	4.9		7.4	0.085	4.9
Var	428.1	0.007	24.0		55.4	0.007	23.8
Ratio of variance (V_A/V_B)				Raw data	7.7	7.7	
				Transformed	1.0	1.0	

The basic principle behind transformation is to bring the data to normality (reduce variation). For examples, Tables 4.8–4.10 show this. The ratio of variance (V_A/V_B) is reduced from 7.7 to 3.0 times when they were transformed using square root. The ratio was further reduced to 1.1 by Log or L_n transformation, and the variance became equal when arcsine or radians transformations were performed, which may indicate that this is the best method unless there is specific need for other transformations.

Once a parametric test is carried out using transformed data, they have to be converted to original scale in order to be presented in the results of the report, papers, or thesis. For example, square root transformed data should be squared $(\sqrt{X})^2$ and log/L_n transformed data should be transformed back to original data using antilog (Log_x). In Microsoft Excel, they can be converted by using = Power (10, log_x) and =Power (2.7148282, Ln_x) functions. Similarly, ArcSine or angular transformation or Square root – ArcSine should be reconverted using $Sin(\sqrt{X})^2$.

Many researchers rush to transform their data when they see in percentage form. However, as a general rule, if the percentage data are within the range of 30–70%, they do not need to transform. If there is even only one observation outside this range, then all of the data set need to be transformed, dividing by 100 before transformation. For example, square root transformation is performed by using $\sqrt{(X/100)}$.

4.8   Questions

Q1. What do you mean by significant figures, and why are they important?
Q2. How do you know whether a particular set of data is in normal distribution?
Q3. How can you avoid the errors in data?
Q4. Why is exploratory data analysis important?
Q5. Why is data transformation done?

4.9   Practical exercises

Ex. 1.
1. Calculate the mean values and standard deviation, then see whether 95% of the data are within the range of mean $\pm$ SD $\times$ 1.96.
2. If a coastal district has 20 rich fishermen's families, 200 middle class, and 80 poor families, what are the probabilities of sampling:
   a) poor and rich families?
   b) poor or middle class families?

Ex. 2. Suppose you have the following data: 4, 5, 6, 7, and 8.
1. Are these discrete variables or continuous? if these values are for:
   a) farm size
   b) family size

c) pH values
d) numbers of leaves in each plant
e) values for DO
f) weights of fish
g) levels used for bad, fair, good, very good, and excellent, respectively.

2. What are the differences between 0 and 0.0, and 100 and 100.000?
3. Round the numbers 106.55, 0.04819, 3.0495, 7815.01, 12.9149, and 20.1500 to
   a) one decimal place
   b) three significant figures
   c) If these were raw data collected by your staff for the pH of your media in your lab experiment, what would you think and how precisely would you present them?
   d) If these were raw data (g) collected by your staff for the weights of individual fish fingerlings, what would you think and how precisely should you present them?
4. Suppose you are designing a new efficient pumping machine. You expect that its capacity is 10 liters per second. If you think it would vary by 1 liter, how precisely should you measure the data to test its efficiency and why?

Ex. 3. In a nursing experiment of fish fry with and without added vitamin mixture in feed, mortalities (%) recorded in each tank are given in Table 4.11.
1. Find the probabilities of keeping fry mortality less than 10% in both groups.
2. Sort the data and present them in a frequency table, bar diagram/histogram, and frequency curve. Compare between the two groups based on the diagrams.
3. Calculate skewness and kurtosis for both groups.
4. Transform data into square root, logarithmic and arcsine forms, draw frequency curve for these transformed data, and compare between the two groups based on the original and transformed data.

Table 4.11 Mortality rate (%) of fish fry in a trial with Vitamin mixture in feed.

Group A: Control									
16.7	13.5	14.3	16.2	14.6	11.9	13.3	15.2	10.5	14.1
12.2	12.9	14.8	13.9	11.6	15.1	14.3	15.2	21.8	15.8
13.8	14.6	25.2	13.7	50.0	12.4	24.7	11.8	11.3	31.1
31.1	11.9	14.4	10.0	10.8	15.8	11.9	10.2	10.9	14.6

Group B: Vitamin- Treated									
9.2	10.9	11.8	11.9	11.6	11.1	12.3	12.2	11.8	11.8
10.1	3.9	10.4	10.7	10.8	0.0	11.9	10.2	10.9	11.6
6.0	11.5	12.3	11.2	11.6	11.9	10.3	11.2	10.5	11.1
10.8	11.6	10.4	10.7	12.0	12.4	11.7	11.8	11.3	1.1

Ex. 4. An experiment was performed to determine the effects of feeding in prawn growth. The two treatments were replicated three times. Freshwater prawns, obtained from a commercial farm, were stocked at 2 prawns $\cdot m^{-2}$ and harvested after 85 days. At harvest, all prawns were counted and individually weighed. Table 4.12 shows the individual weight of harvested prawns. Take a look at the data table provided carefully, and as a student of statistics:

1. Point out the extreme values or odd things, including consistency in significant numbers as well as formats that you would like to correct them.
2. Round the numbers (those that need rounding), to suitable significant figures.
3. Plot scattered graphs, draw line graphs, and make bar charts for each data set. Point out the errors/extreme values, and then double check the data against original data to correct them (original will be provided later).
4. Calculate means for each data set before and after correction. Then, compare them and explain how the conclusions and recommendations could be affected by mistakes on data entry.
5. Using Microsoft Excel or any statistical package, group the data and present them in a frequency table, bar diagram, histogram, and frequency polygon. Compare the treatments based on these diagrams.

Table 4.12 Individual weight (g) of freshwater prawn at harvest.

S.N.	Fertilization Only			Fertilization with Feeding		
	Rep. 1	Rep. 2	Rep. 3	Rep. 1	Rep. 2	Rep. 3
1	100.0	8.02	112.4	14.1	18.44	22.0
2	5.4	8.2	28.1	15.5	15.0	12.6
3	5.9	8.7	6.3	16.0	15.4	15.0
4	9.3	9.9	8.778	17.5	15.5	15.112
5	9.4	9.9	10.4	18.1	16.4	16.6
6	9.9	10.0	14.4	20.9	17.2	16.9
7	5.0	11.1	14.5	21.1	17.8	17.1
8	10.0	11.6	15.8	22.3	18.4	17.8
9	100.1	12.0	16.2	23.3	20.5	18.0
10	**10.6**	**12.0**	**16.4**	**23.6**	**21.1**	**19.4**
11	10.7	12.5	16.5	24.0	22.3	22.3
12	13.1	12.6	19.2	28.8	25.1	23.69
13	13.2	13.5	24.8	31.0	25.5	24.2
14	13.6	14.7	17.2	35.77	26.3	26.7
15	14.4	16.7	11.4	35.8	27.7	27.1
16	14.5	16.7	13.9	17.0	28.1	27.4
17	15.3	16.9	10.4	17.44	31.6	28.8
18	15.4	16.9	11.6	17.4	32.1	30.7
19	15.7	17.4	11.8	20.0	32.7	13.22
21	16.4	18.6	14.4	21.8	33.2	14.5
21	17.9	19.9	12.5	22.6	34.4	16.0
22	18.6	20.6	12.7	27.4	34.6	18.5
23	20.0	20.6	12.7	28.6	36.3	18.9
24	20.8	22.02	112.9	18.9	37.5	19.5
25	22.67	23.2	12.9	18.9	9.9	19.7
26	35.5	24.4	14.4	19.9	13.222	20.6
27	8.0	24.5	14.8	20.1	15.7	20.6
28	8.0	24.6	15.8	20.4	17.8	20.7
29	8.6	26.9	15.9	20.6	20.3	22.0
30	8.7	30.3	17.3	22.4	20.3	15.5
31	9.2	10.6	18.5	23.1	22.0	17.3
32	9.6	12.0	19.2	24.6	22.2	18.111
33	10.4	12.6	15.8	24.9	23.2	18.2
34	10.4	12.9	16.7	25.1	18.1	19.9
35	111.7	13.1	17.7	26.2	20.5	21.2
36	12.7	13.6	20.1	26.3	21.0	21.5
37	0.6	*12.1*	*20.9*	21.9	21.1	
37	10.5	12.4	14.3	22.1	21.9	
39	10.6	14.3	14.4	29.2	22.0	
40	11.2	14.5	14.7		22.4	
41	*13.9*	*14.7*	*14.9*		23.2	
42	14.4	15.0	15.1		25.8	
43	14.7	15.333	15.3		26.0	
44	14.7	117.1	15.5		27.8	
45	14.8	18.6	16.8		29.6	
46	15.0	18.7	17.7		30.1	
47	15.9	18.7	17.0			
48		19.3	19.3			
49		18.5	18.9			
50						

Chapter 5

Central locations and variability

5.1   Concept and importance

Central tendency is a single value that represents or describes the whole population or a sample of particular characteristics. In other words, it is the measure of a central location or the value or characteristics that fall in or near the middle. It was originated from the concept of "average man," based on which people make up their minds on similarities or differences between or among particular groups. An example is the average height or IQ of Asian students as compared with Japanese or American students. Similarly, if we want to know the productivity of fish per hectare in a pond system in Thailand, we would think average, which is 4 t·ha^{-1}; that means that about half of the farmers get less and the other half achieve higher than that. However, when we need to state the productivity of pond aquaculture, we use the average value to represent the whole country. Such representative values are widely used as bases while planning research projects or development programs. Ideally, a representative value should fall in the middle, as shown in Figure 5.1, but in a real world situation, it may not. In some cases, a single value may not be able to represent the whole data set. Therefore, a number of ways have been developed for the measurement of central representative value(s), which are described in the following section.

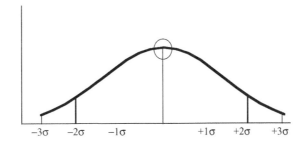

Figure 5.1 In a normally distributed data, most data points are in the middle.

5.2   Measure of central tendency

Although the most common central location is the average or mean (arithmetic, geometric, and harmonic), there are also other types of central locations, i.e. median, mode, midrange, midhinge, quartiles, and percentiles. Their use depends on the nature of data and propose of the research. For a normally distributed data set, arithmetic mean can represent the central point; whereas in other cases, or if data are not normally distributed, median and mode can be more useful and appropriate. In some cases, other central points and subcentral points, e.g. midrange, midhinge, quartiles, and percentiles, might provide better and/or additional information about further distribution and cluster of observations. This section discusses their computation methods and their usefulness.

5.2.1   Mean

5.2.1.1   Arithmetic mean

Arithmetic mean (AM) or average is the most commonly used value to describe the central point or to measure central tendency. It is often referred to as the center of gravity of the data. AM is calculated by adding the values of all of the observations, i.e. sum (Σ), and dividing by the number of observations (n). AM represents the data very well if they are in normal distribution or if the observations follow the arithmetic series, e.g. 2, 4, 6, 8…($X + 2$). Therefore, the following observations can be estimated by adding a constant value to its preceding ones. From the definition,

$$\text{Arithmetic mean} = \frac{\text{Sum of all observations } (\Sigma X)}{\text{Total observations } (n)}$$

mathematically, AM (average) is expressed as:

$$\bar{X} = \frac{X_1 + X_2 + X_3 + \cdots + X_n}{n}$$

$$\bar{X} = \frac{\sum\limits_{i=1}^{n} X_i}{n}$$

or if some of the observations are repeated, then AM can be arranged along with their respective frequencies, and the equation becomes:

$$\bar{X} = X_1 f_1 + X_2 f_2 + X_3 f_3 + \cdots + X_n f_n$$

$$\bar{X} = \frac{\sum\limits_{i=1}^{n} X_i f_i}{n}$$

For example, what is the mean size of fish if we have recorded 10 g, 12 g, 15 g, and 20 g for individual weight of four random samples drawn from a tank after a 30-day nursing trial?
Here,

$$\text{Sum}(\Sigma) = 10 + 12 + 15 + 20 = 57$$

Number of observations $(n) = 4$
Therefore, average or mean $(\bar{X}) = (10 + 12 + 15 + 20)/4 = 57/4 = 14.25$ g.
While calculating the mean, in some cases, an investigator has to take the relative importance of each observation into account. The difference in importance (weight) may be due to a different number of observations for particular figures or the values may carry different weights/ranks. In such a case, AM is often referred to as weighted mean/average:

$$\text{Weighted mean} = \frac{\Sigma(\text{observed values} \times \text{weights of observations})}{\text{Total weights}}$$

which is expressed as:

$$\bar{X} = \Sigma(Xw)/\Sigma w$$

The example in Table 5.1 shows the scores of two students for aquaculture entrance using weights for different subjects based on their relevance. The score in biology gets 3 weights because it is the most important subject for aquaculture, and other subjects get lower scores because they are of less importance to the study of aquaculture. In this case, although Student 2 has a higher total score (210 vs. 215), Student 1 has higher weighted scores (total and weighted mean) because of the higher weight given to biology, which is the most important background knowledge required for aquaculture. Because Student 1 received a higher score in biology, s/he will be preferred over Student 2.

Table 5.1 Comparison of two students based on the simple score and weighted scores.

Subjects	Score (%)		Weights (w)	Weighted Score (Xw)	
	Student 1	Student 2		Student 1	Student 2
English	60	80	2	120	160
Mathematics	70	75	1	70	75
Biology	80	60	3	240	180
Total	210	215	6	430	410
Mean score	$210 \div 3 = 70$	$215 \div 3 = 71.6$			
Weighted mean score	$430 \div 6 = 71.7$	$410 \div 6 = 68.3$			

5.2.1.2   Geometric mean

If data are in geometric series (multiplication of a constant value), e.g. 2, 4, 8, 16... $(X \times 2)$, then geometric mean (GM) represents the central location. It is calculated by multiplying the values of all the observations and nth root, such as:

$$GM = \sqrt{X_1 \cdot X_2 \cdot X_3 \ldots X_n}$$

If data are in a geometric series, log transformation is required before computing the GM. After computing the GM, it is then transformed back to the actual GM.

$$GM = \text{antilog}[1/n \; \Sigma \; (\log_X)]$$

5.2.1.3   Harmonic mean

Harmonic series is the reciprocal of arithmetic series; for example, if the arithmetic series is 2, 4, 6, 8 then the harmonic series is 1/2, 1/4, 1/6, 1/8, which is expressed as $1/(x + 2)$. Likewise, harmonic mean (HM) is the reciprocal of the arithmetic mean, which is the sum of the reciprocals of the observations divided by the number of observations (items):

$$HM = 1/n \; \Sigma \; 1/X$$

HM is used in the engineering field for computing average speed, in microbiology when concentrations are expressed as reciprocals, and in post-harvest technology for the color intensity of the product, and so on.

5.2.2   Median

AM can be used only if data are normally distributed. In some cases, AM does not represent the data, e.g. per capita income of the countries, salaries of staff in most organization, and other cases in which data vary a lot. In such cases, standard deviations are higher than the means. It is necessary to find a value that can represent this type of data set at least somehow and some parts, if not whole data. The median is one of the most popular locations after mean to represent data.

After sorting and arranging a data set as an array, the value that falls right in the middle of the scale is called the median, which divides the values in such a way that 50% of the observations are smaller and 50% of the observations are larger.

Table 5.2 shows an example in which AM may not represent the salary of all the staff of a fish farm. It shows the salaries of 21 staff members, and salaries range from $109 to $1,515. In this case, the calculated AM ($287) does not correctly represent all of the staff; the standard deviation ($332) is higher than

Table 5.2 Salary of staff of a fish farm in Thailand.

SN	Position	Monthly Salary (US$)
1	General Manager	1,515
2	Assistant manager	909
3	Hatchery manager	364
4	Grow-out manager	333
5	Marketing manager	333
6	Accountant	273
7	Technician 1	242
8	Technician 2	212
9	Technician 3	197
10	Technician 4	197
11	Technician 5	197
12	Labourer 1	139
13	Labourer 2	136
14	Labourer 3	133
15	Labourer 4	130
16	Labourer 5	127
17	Labourer 6	127
18	Labourer 7	127
19	Labourer 8	109
20	Labourer 9	109
21	Labourer 10	109
	Total	6,021
	Mean	287
	SD	332
	Max	1,515
	Min	109

the average, and only 5 staff members make above the average, whereas the other 16 make salaries below the average.

In this case, instead of AM, other types of central locations would be appropriate to describe the data. An example is median value, which can be located by arranging data in ascending or descending order. If there is an odd number of samples or observations, middle value can be easily pointed out. For example, $197 is the median salary of fish farm staff (Table 5.2) as it falls on the 11th position. However, if the number of observations or samples is even, the median value is the sum of the 2 middle values divided by 2. The following examples should make this clear.

Case 1: Odd number of observations.
Raw data: 3, 2, 4, 5, 10, 8, 9, 6, 7
Data array (sorted) = 2, 3, 4, 5 ⑥,7, 8, 9, 10
Here,
Number of observation $(n) = 9$
Median $= [(n + 1) \div 2]^{th}$ value $= 10/2$
        $= 5th$ value in the data array, i.e. 6

Table 5.3 A hypothetical data set with frequencies.

Observation (cm, x_i)	Frequency (f_i)	Cumulative Frequency	$x_i f_i$
3.3	1	1	3.3
3.4	0	1	0
3.5	1	2	3.5
3.6	2	4	7.2
3.7	1	5	3.7
3.8	3	8	11.4
3.9	3	11	11.7
4.0	4	15	16.6
4.1	3	18	12.3
4.2	2	20	8.4
4.3	2	22	8.6
4.4	1	23	4.4
4.5	1	24	4.5
Total	24		95.0

Case 2: Even number of observations.

Data array $= 2, 3, 4, 5, 6, \boxed{7, 8,}\ 9, 10, 11, 13, 14$
Total observations $= 12$
Median $=$ (6th value + 7th value)/2
$\qquad = (7 + 8)/2 = 7.5$

Case 3: Table 5.3 presents a data series with frequencies.
Here,
Total number of observations (n) $= 24$
Median $=$ (12th + 13th values)/2 $= 4.0$ cm
If 4.0 represents a class, it is necessary to locate the median within the class. As class interval is equally spaced, it can be divided by the number of frequencies that fall in that class, e.g. 4 observations are in the class 4.0 or (3.95–4.05). The class 4.0 can mean (3.95–4.05).

$$= \text{Lower limit} + [(4.05 - 3.95)/4]$$
$$= 3.95 + 0.025$$
$$= 3.975 \text{ cm}$$

Therefore, median $= 3.975$ cm

5.2.3 Mode

Mode is defined as the value that appears most frequently in a given set of data. It can also be considered a "typical" item that carries important information. In

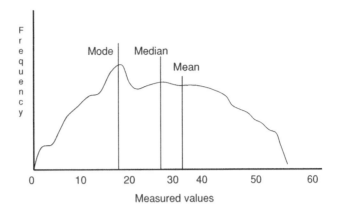

Figure 5.2 Location of mode as compared with mean and median.

many cases, mean and median are not adequate to describe the data, especially if there are extreme values and high variability. In such cases, locating the most common or clustered observations, i.e. the modes, is more appropriate, no matter whether a few observations are far away from that location. In an ungrouped data set, a mode is the actual value measured or recorded, but in a grouped frequency distribution, the mode refers to the modal class rather than one particular value. In a perfectly normal data set, mode is close to the middle; however, depending on the nature of variables, it can be anywhere (see Figure 5.2). A set of data can have more than one mode. If a data set has two values appearing most often, the data set is said to be "bimodal" (Figure 5.3). A data set can have even more than two modes; this is called a multimodal data set. On the other hand, a data set may have no mode. If a frequency distribution is U-shaped, which means the value in the middle expected to be the highest in frequency appears instead to be the lowest, then the lowest value is referred to as the antimode.

In some cases, mode can describe a data set more appropriately. For example, using a data set of per capita income (Table 5.4), if we report based on the mean, we have to say that mean per capita income of 19 Asian countries is \$5,802; however, this does not make sense because no country has close to the mean and there is very high variability among the countries. Similarly, if we say the median per capita income is \$1,080 (e.g. the Philippines), this is more representative for

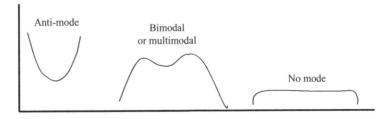

Figure 5.3 Distributions showing animode, two modes, and no clear mode.

Table 5.4 Per capita income of 19 Asian countries.

Group	Per Capita (US$)	Income Group	Frequency (no. of countries)
1. Japan	34,510	>10,0000	4
2. Hong Kong	25,430	2,000–10,000	4
3. Singapore	21,230	1,000–2,000	2
4. S. Korea	12,030	500–1,000	3
5. Malaysia	3,780	<500	6
6. Maldives	2,300		
7. Thailand	2,190	**Total**	**19**
8. Iran	2,000		
9. China	1,100		
10. Philippines	1,080		
11. Sri Lanka	930		
12. Bhutan	660		
13. India	530		
14. Mongolia	480		
15. Vietnam	480		
16. Pakistan	470		
17. Bangladesh	400		
18. Lao PDR	320		
19. Cambodia	310		
Mean	**5,802**		
SD	**10,075**		

many countries, e.g. from Sri Lanka to Malaysia, but it still misleads the readers because there is no way that 4 of the top countries and 8 of the bottom can be represented by that median value. In this case, using the idea of frequency distribution after grouping and locating a mode would be more appropriate. For example, grouped data showed that, although few Asian countries (4) in the list have very high income ($10,000), many of them (6) have lower than $500 per capita income. This is certainly more valid and appropriate reporting.

An example of modes can be explained using the model size of fish farms, e.g. the most common size of farm that the majority of fish farmers have in any country/region. The majority of farmers may have a 2-ha fish farm in most countries, although they may range from 0.2 ha to 500 ha (high variation or SD). As few farmers have very big farms, the average farm size can be larger than 2 ha. If the majority of farmers have 2-ha fish farms, then there must be a reason. For example, government policy or promotion might have recommended that size. Discovering the reason is one of the more important aspects of the research.

5.2.4   Midrange and midhinge

In some cases, the average of the smallest and largest observations, i.e. the midrange, can be useful. In other cases, the average of the first and third quartiles might be used as a central location, which is called midhinge. These are expressed as:

$$\text{Midrange} = (X_{\text{smallest}} + X_{\text{largest}})/2$$

$$\text{Midhinge} = (Q_1 + Q_3)/2$$

5.2.5   Quartiles, percentiles, and others

In a data set, there are other important points to locate. Various ways of locating these central points of such subgroups have been used. The most commonly used are quartiles, deciles, percentiles, and so on.

Quartiles are the values or observations that divide the whole set of ordered data into four equal parts. Therefore, there will be three cut-off points: the first, second, and third quartiles. The second quartile lies right in the middle; therefore, it is actually the median. They can also be defined in another way. In an ordered data set, if the value has 25% of the smaller observations on one side and 75% on the other side, it is called the first quartile (Q_1). The value that has 75% smaller values on one side and the remaining 25% on the other side is called the third quartile (Q_3). Figure 5.4 shows the locations of quartiles, and the Box-and-Whisker plots (bottom of Figure 5.4) are used to show this in a more attractive way.

The positions of the quartiles can be located as:

$$Q_1 = \frac{n+1}{4} \text{ ordered observation} \quad Q_3 = \frac{3(n+1)}{4} \text{ ordered observation}$$
$$Q_2 \text{ or median} = (Q_1 + Q_3)/2$$

Similarly, percentiles and other fractiles can be used whenever necessary. For examples, quintiles divide a data set into 5 equal parts, deciles divide 10 equal parts, and percentiles divide 100 equal parts. In percentile scale, the 25th percentile means the first quartile, 50th percentile is the median, and 75th percentile

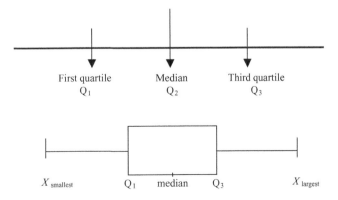

Figure 5.4 Box-and-Whisker plots that uses quartiles as basis.

is the third quartile. It is easier to find the value of the required fractile item in a grouped frequency distribution. The location of the fractile item can be determined by multiplying a fraction by the total number of observations. For example, if a data set has 36 observations,

Case 1: the third quartile of the data $= 3/4 \times 36 = 27$th observation; and

Case 2: the 60th percentile of the data $= 60/100 \times 36 = 21.6 = 22$nd observation.

The application of fractiles can be found in medical research. For example, LC50 is the lethal concentration (ppm, etc.) of certain medicines at which 50% of the animals die in an hour or certain time. LD50 is the lethal dose at which 50% of organisms die in a certain period of time. ED50 means effective dose at which 50% animals are cured. In such cases, death of 50% organisms is adequate to see the efficacy and the rest is not necessary. Similarly, percentiles are used as cut-off values. For example, in a normal distribution, lower than 2.5% and above 97.5% values of the distribution are considered extreme values. The investigator is interested in only the middle 95% of values without considering the first and the last 2.5% values in the tail area.

5.3   Measure of variability

Variability implies how the observations are either scattered all over or clustered around the central location. Variability is the basis for comparison, without which the definition of statistics is incomplete. As an example, statistics has been defined as "the scientific study of numerical data based on variation in nature." Similarly, it has also been defined as "science of analyzing data and drawing conclusions, taking variation into account." The variability or dispersion is measured by using various parameters, which are described in the following sections.

5.3.1   Range

Range is the difference between the largest and the smallest observations in a set of data.

$$\text{Range} = |X_{\text{smallest}} - X_{\text{largest}}|$$

It is the simplest and a very crude measure of dispersion. It considers only how far the two extreme values are from the center; therefore, it is very much affected by a single extreme/outlying observation and sample size. It doesn't take into account how and where other observations are clustered or dispersed. When expressing the range in writing, the lowest and highest observations or values themselves are shown rather than the calculated difference between them. For example, "the average survival of prawn in an experiment was 85% (range 65–94%)."

5.3.2   Interquartile range or quartile deviation

Interquartile range is the difference between the third and the first quartiles, i.e. $Q_3 - Q_1$. It considers only the central half of the data set; therefore, it is not affected by the extreme values or outliers. It sounds better than range in that sense, but it ignores the values below the first quartile and beyond the third quartile. Half of the interquartile range is called quartile deviation.

Quartile deviation $= (Q_3 - Q_1)/2$

5.3.3   Mean deviation

With mean deviation (MD), dispersion of data is measured more comprehensively considering all the deviations of observations from the central location. The average of these deviations is the MD or average deviation, which is expressed as:

$$MD = \frac{\sum (Y_i - \bar{Y})}{n}$$

But, in a normal population, 50% of observations are higher than the mean and 50% are lower, so the sum of these deviations is zero (0). Therefore, statisticians started to use absolute differences or deviations from the mean, ignoring $+/-$ signs.

$$MD = \frac{\sum |Y_i - \bar{Y}|}{n}$$

Another way to eliminate the negative sign is to square the deviations:

$$MD = \frac{\sum (Y_i - \bar{Y})^2}{n}$$

5.3.4   Variance and standard deviation

MD was popular during the early 20th century, but now variance (Var) is widely used and has become the fundamental basis for analysis as it is used for probability and hypothesis testing. The average of the squared deviations is variance. As the variance is a squared unit, e.g. m^2, positive square root of variance, which is called standard deviation (SD), is used for the presentation purpose to express variation of a particular mean as:

$$SD = \sqrt{\sum (Y_i - \bar{Y})^2 / n}$$

5.3.5   Population and sample variance/SD

If data are collected from samples for analysis, calculations of variance and SD using the aforementioned equations are underestimated and biased. In order to correct this, the deviations are divided by degree of freedom $(n-1)$ instead of the total number of observations (n). Therefore, these variability parameters are estimated as shown below:

$$\text{Population SD (s)} = \sqrt{\left[\frac{\sum_{i-1}^{N}(X_i - m)}{N}\right]^2}$$

$$\text{Sample SD (s)} = \sqrt{\left[\frac{\sum_{i-1}^{n}(X_i - \bar{X})}{n-1}\right]^2}$$

Variance and SD can be computed using raw data without calculating a mean using the following equations:

$$\text{Variance } (s^2) = [\Sigma X_i^2 - (\Sigma X_i)^2/n]/(n-1)$$

$$\text{SD}(s) = \sqrt{[\Sigma X_i^2 - (\Sigma X_i)^2/n]/(n-1)}$$

This method helps reduce efforts and time. For example, we will calculate the variance or SD of the following data set: 1.2, 1.4, 1.6, 1.8, 2.0, 2.2, and 2.4. Table 5.5 shows the way these methods can be used:
Here,
Number of observations $(n) = 7$
Range $= 2.4 - 1.2$
Mean deviation $= 2.4/7 = 0.34$ g

Table 5.5 Calculating SD by using the mean.

Data Y (g)	$(Y - \bar{Y})$	$/Y - \bar{Y}/$	$(Y - \bar{Y})^2$
1.2	−0.6	0.6	0.36
1.4	−0.4	0.4	0.16
1.6	−0.2	0.2	0.04
1.8	0.0	0.0	0.00
2.0	+0.2	0.2	0.04
2.2	+0.4	0.4	0.16
2.4	+0.6	0.6	0.36
$\Sigma Y = 12.6$	0.0	2.4	1.12
Mean $(\bar{Y}) = 1.8$	–	0.34	–

Table 5.6 Observations and frequencies.

X (g)	X² (g²)
1.2	1.44
1.4	1.96
1.6	2.56
1.8	3.24
2.0	4.00
2.2	4.84
2.4	5.76
$\Sigma X = 12.6$	23.80

Variance $(s^2) = \Sigma(Y_i - \bar{Y})^2/(n-1) = 1.12/6 = 0.187$ g², SD $= 0.43$ g
Table 5.6 describes the calculation of SD directly from raw data.
Here,
Number of observations $(n) = 7$
Range $= 2.4 - 1.2 = 1.2$ g
$s^2 = [\Sigma(X_i^2) - (\Sigma X_i)^2/n]/(n-1)$
$\quad = [23.8 - (12.6)^2/7]/6 = 1.12/6$
$\quad = 0.187$ g²
SD $(s) = \sqrt{0.187} = 0.43$ g

5.3.6 Standard error

Standard error (SE) has become popular recently. Researchers often misunderstand and misuse it. Variability of observations within a data set is SD, whereas variability of two or more means is SE. Therefore, it is often called standard error of means. It is computed as SD of two or more means divided by the square root of the number of observations: SE $=$ SD $\div \sqrt{n}$.

Most researchers may incorrectly calculate it as the average of SDs. Table 5.7 shows how to compute these.

Table 5.7 Sample means, treatment means, SDs, and SEs.

Treatment (T1)	Sample A	Sample B	Sample C
R1	10.1	12.0	12.2
R2	10.0	14.3	11.1
R3	11.3	11.2	13.1
R4	12.1	10.5	16.3
Total	43.5	48.0	52.7
Replicate mean	**10.9**	**12.0**	**13.2**
Standard deviation (SD)	**1.0**	**1.7**	**2.2**
Treatment mean	$(10.9 + 12.0 + 13.2)/3 = 12.02$		
SE	=STDEV(10.9, 12.0, 13.2)/$\sqrt{3}$ = 0.66		

Note: =STDEV(...) is the formula function for SD in Microsoft® Excel.

5.3.7   Coefficient of variation

Coefficient of variation is the percentage measure of variation relative to the magnitude of a mean, which is computed as:

$$\text{Population CV} = \left(\frac{s}{m}\right) \times 100\% \quad \text{and} \quad \text{Sample CV} = \left(\frac{s}{\bar{X}}\right) \times 100\%$$

From the example given in Tables 5.5 and 5.6, the coefficient of variation (CV) = (0.43/1.8) × 100 = 24%.

5.3.8   Implications of variability

A highly dispersed data set has larger range, variance, SD, and SE. In other words, a highly spread out data set is low in precision and accuracy. Conversely, more concentrated, precise, or homogenous data will have smaller range, variance, and SD (with high precision and accuracy). If all the observations are the same, the range, variance, and SD will be zero. None of these measures can be negative. Two distant means with little variations are more likely to be significantly different, and means with high variations are likely not to be significantly different.

In Figure 5.5, there is a higher chance of a significant difference between two means in Group A (very little overlap), whereas there is no chance of significant difference in Group B as the overlap seems to be more than 5%.

From a statistical point of view, presenting means without variability (SD or SE) has no meaning. Either SD or SE needs to be presented along with the means

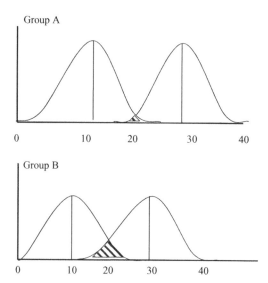

Figure 5.5 Frequency distributions with low and high overlaps.

Table 5.8 Means for Treatments A and B (mean ± SE).

Treatments	Mean
Treatment A	24.6 ± 4.2 (SD)
Treatment B	30.2 ± 3.3 (SD)
Overall mean	27.4 ± 2.8 (SE)

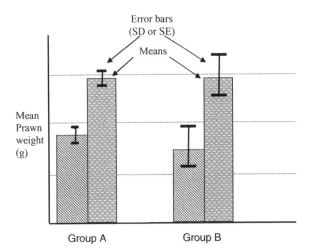

Figure 5.6 Graphical presentation of SD or SE is shown using error bars.

while presenting results both in tabular and graphical forms; mean ± SD or SE in tabular form is shown in Table 5.8, and error bars in graphical presentations are show in Figure 5.6. Both of them show that a mean can go up and down by that amount. Therefore, most researchers have started using error bars as they show variations very clearly. For example, in Figure 5.6, it can be seen that Group B has high variation, whereas in Group A, both treatments have less variability. More importantly, error bars drawn using SE or SD may serve as rough indicators of the presence or absence of significant differences between means. If the error bars of two means are overlapping, it is almost sure that they are not significantly different; if they are not overlapping with large distance, there is a high chance of a statistically significant difference. However, this has to be confirmed by using the appropriate statistical tool.

5.4   Questions

Q1. Why are central locations of data sets important?

Q2. Explain why mean without standard deviation or standard error has no meaning.

Q3. Why is median important?

Q4. What is the use of mode?

5.5 Practical exercises

Ex. 1. Calculate the following parameters for each replicate of the experimental data using Microsoft® Excel (formulas and data analysis functions) as well as other statistical packages.
Central tendiency/locations:

- Mean (arithmetic, geometric, harmonic and weighted means)
- Median
- Mode
- Quartiles (Q_1, Q_2, and Q_3)
- Percentiles (5%, 25%, 33%, 50%, 66%, 75%, and 95%)

Dispersion/variability:

- Minimum (Min)
- Maximum (Max)
- Range
- Midrange
- Midhinge (you can make a Box-and-Whisker Plot using SPSS®)
- Variance (Var)
- Standard deviation (SD)
- Standard error (SE)
- Coefficient of variation

Ex. 2. Solve the following problem.
Suppose you had stocked 720 broodfish (1:1 sex ratio) in 20 jumbo hapas. After 3 months, the data set shown in Table 5.9 was recorded by your staff at the end of the experiment. You were supposed to compare the survival rate between male and female broodfish. After taking a look at the data sheet, you have noticed that the 19th hapa (*) has no data recorded. No special note from your staff was found about this hapa. How would you compare the survival?
Hints: There are three main possibilities:

1. As there are very low survivals of males in the 19th hapa, you could imagine that all the females might have died, i.e. 0 females.

Table 5.9 Number of males and females counted in each hapa at the final harvest.

Hapa No.	1	2	3	4	5	6	7	8	9	10
Male	345	360	350	345	322	330	340	366	324	306
Female	320	320	200	301	322	315	300	200	200	290
Hapa No.	11	12	13	14	15	16	17	18	19	20
Male	298	265	234	350	258	321	310	240	126	275
Female	220	322	320	222	210	242	280	233	*	230

2. It could be an accident which might have resulted in escaping of those females and mixed with other fish before counting. This can be considered as data not available (N/A) or missing values.

3. Sometimes we can't analyze even if we lose only a single datum. In such a case, we have to estimate the missing value using a statistical formula:

   Missing value = [(b × block total) + (t × treatment total)
   – grand total]/(t − 1)(b − 1)

   where b is the number of blocks and t is the number of treatments.

Chapter 6
Basics of hypothesis formulation and testing

6.1 Concept

A hypothesis is an assumption made for the sake of argument and claim after having made observations of the natural phenomena. It is the starting point of scientific discovery and invention. Normally, it is testable and provides a possible explanation of a certain phenomenon or event. A hypothesis can be thought of as an embryo, which might develop into a theory and then become a law. If a hypothesis is not testable, it implies insufficient evidence to provide more than a tentative explanation, e.g. hypotheses to explain the extinction of dinosaurs, the origin of the universe, or extinction of sea food species. Whereas theory implies a greater range of evidence or greater likelihood of truth. The theory of evolution and the "law" imply a statement of order and relation in nature that has been found to be invariable under the same conditions, e.g. the law of gravitation, law of inheritance, and so on.

Any new knowledge or belief has to be tested or compared against the existing one. Therefore, while comparing between them, a null hypothesis (H_0) is established that assumes there is no difference between the new and old ones. However, there can be some indications that existing knowledge or beliefs may not be true, and an alternate idea might prove to be true. The new idea is called the alternate hypothesis (H_A) and assumes that the new idea is better or true and goes against the traditional belief. A hypothesis (alternative) is sometimes known as "an intelligent guess" based on limited information. Although experimental results may match predictions, no one should believe an "intelligent guess" is true. Without proper testing and analysis using appropriate tool(s), which includes designing an experiment or survey to generate or collect data/information (raw), exploratory data analysis, choosing an appropriate significance level or confidence limits (intervals), and analyzing data by using appropriate statistical tools. In statistical procedure, null hypothesis is tested, not the alternative hypothesis. Based on the data outputs or results, the null hypothesis is either accepted or rejected, which is expressed mathematically as:

$H_0 = H_A$ which means there is no difference $\rightarrow$ Accept H_0

$H_0 \neq H_A$ which means there is a difference $\rightarrow$ Reject H_0

As the hypothesis is either accepted or rejected based on the numerical facts, only trustworthy data can be used for this purpose. Hypothesis testing therefore involves data collection using proper methods, compilation, and securitization, use of appropriate tools for analysis, and judicious decision, interpretation, and explanation.

6.2 Significance level

Probability (P) of occurrence of any event by chance or random error is popularly known as significance level. The level of error can be higher or lower depending on the situation. For example, in very controlled laboratory research, chance of error can be 1% or even less. Therefore, level of significance is considered 1% for those cases; whereas, in aquaculture and other field-based biological research, 5% is considered a significant level. In social survey research, the significant level is normally considered 10%. In fact, there is no fixed level of significance. Researchers themselves determine the significance level depending on the conditions. For example, a 40% level of confidence may be considered adequate if a drug is found to be effective (i.e. 60%) against AIDS since there is no effective drug invented so far. On the other hand, concerning drugs for the common cold, a probability as low as 0.01% may be necessary to convince that the treatment does not cause any side effects. Therefore, ultimately, the researchers are the ones to fix the significance level rather than depending on a machine or mindless mechanistic statistical significance.

6.3 Confidence level, limits, and interval

When concluding that any hypothesis is true or false, there is a certain level of confidence. From a statistical point of view, nothing is absolutely true or absolutely false. Therefore, 100% confidence is very rare. In social survey research, a 90% confidence level might be enough; however, in most biological research, a confidence level of 95% is considered sufficient, and in medical research, the confidence level is usually as high as 99%.

Any mean has two confidence limits: the lower limit (LL) and the upper limit (UL) for a given level of confidence. The difference between the two limits is called confidence interval (CI). The sample mean estimates the true mean, and standard error (SE) describes the variability of that estimation. This variability can be conveniently expressed in terms of probabilities by calculating CIs. The following example describes these methods.

Suppose a sample of 40 fish was drawn from a pond containing 3,000 fish. If the computed mean is 100 g and standard deviation (SD) is 35 g, what are the

LL and UL for 68%, 95%, and 99% CIs? For this, SE is computed as:

$$SE = SD/\sqrt{n} = 35/\sqrt{40} = 5.5 \text{ g}$$

The range mean $\pm$ 1 SE covers 68% in the normal curve. Or by simply adding/subtracting 1 SE to/from the mean, we can get the LL and UL for 68% CI:

$$LL = 100 - 5.5 = 94.5 \text{ g}$$
$$UL = 100 + 5.5 = 105.5 \text{ g}$$
$$CI = 105.5 - 94.5 = 11 \text{ g}$$

However, for 95% and 99% CIs, SE has to be multiplied by *t*-statistics, which depend on the degree of freedom (df), before adding to or subtracting from the mean (the critical value for 39 df of the *t*-distribution):

$$t\text{-statistics for 95\% CI}$$
$$= (t_{0.05,\ 39}) \times \text{ SE for } n - 1 = 39$$
$$= 2.023 \times 5.5$$
$$= 11.1 \text{ g}$$
$$LL = 100 - 11.1 = 88.9 \text{ g}$$
$$UL = 100 + 11.1 = 111.1 \text{ g}$$
$$CI = 111.1 - 88.9 = 22.2 \text{ g}$$

In this case, we can now say with 95% confidence that the true mean falls between these limits ($88.9 - 111.1$ g). Similarly, CI further increases as we want higher confidence, e.g. for 99%:

$$t\text{-statistics} = \text{mean} \pm (2.708 \times 5.5) = 100 \pm 14.9 \text{ g}$$
$$LL = 100 - 14.9 = 85.1 \text{ g}$$
$$UL = 100 + 14.9 = 114.9 \text{ g}$$
$$CI = 114.9 - 85.1 \text{ or } 14.9 \times 2 = 29.8 \text{ g}$$

This shows that CI increases as the confident level increased. Wider range is necessary to be confident that the sample mean will fall within the range.

6.4   Statistical and biological significance

In some cases, two means may have statistical difference but that may not affect the practical life situation where we apply it. For example, Samples 1 and 2 are the weights of fish in grams:
Sample 1: 100.1, 100.2, 100.3, 100.1, 100.2, 100.3 → Mean 100.2 g.
Sample 2: 100.4, 100.5, 100.5, 100.6, 100.6, 100.4 → Mean 100.5 g.

Student's *t*-test shows that the difference between these two means is highly significant ($P < 0.01$). However, the actual difference ($100.5 - 100.2 = 0.3$ g) has no biological significance because the difference between two fish weighing 100.2 g and 100.5 g would not matter much. However, if we were weighing gold or some other precious substance, a weight difference of 0.3 g would be economically significant and care would be taken while measuring. There is a saying that the difference to be a difference must make a difference. Therefore, a researcher should see whether any statistically significant difference, often called detectable effect size, has any practical concern. In practical life, terms such as concept of biological or economic significance, effect size, or substantive importance/significance/meaningful differences are frequently used. For example, if you feed vitamin C at 50 g·kg^{-1} of feed to your fish, the survival will be increased by 10%. This 10% survival, which is claimed due to the use of vitamin C, is substantial and has an economic value for farmers. However, an increase lower than this may be still statistically significant but may not be considered substantial.

6.5   Errors in hypothesis testing

Proper designing of the research trial or survey and selection of appropriate statistical tools while analyzing data play crucial roles in avoiding or minimizing errors. However, in the real world, errors occur quite often. There are two types of error in hypothesis testing: Type I and Type II. Rejecting H_0 when it is true is Type I error, also called a error. Accepting H_0 when it is wrong is Type II error, which is also called b error. The power of the statistical test is $1 - b$, which means probability of rejecting the H_0 when it is wrong and should be rejected. In other words, it is the probability of making the right decision or detecting significant difference when it exists. Therefore, power analysis is important, especially when the results are nonsignificant. Power analysis can reveal whether the replication was adequate for any treatment to show its effects. A minimum of 80% (or b = 0.20) is considered acceptable statistical power (Searcy-Bernal 1994), the higher the better. Table 6.1 shows a summary of errors.

The two types of errors are often compared. It is better to miss significant difference/relationship if there is one than to claim significant difference when there is none. For example, a researcher found that a new strain of tilapia gave higher production compared with the local strain and recommended that farmers to grow the new one. But, when the second researcher tested later using the

Table 6.1 Types of errors in hypothesis testing.

Decision	When H_0 Is True	When H_0 Is False
H_0 rejected	Type I Error (a)	✓
H_0 accepted	✓	Type II Error (b)

same experimental protocols, he did not actually find any difference. In this case, the first researcher committed the Type I error. In another trial with catfish, the researcher did not see any difference between new and old strains and didn't suggest that farmers use the new strain. But later, another researcher found that the new strain could actually produce more. In this case, statistical difference was not detected, so the first researcher made a Type II error. If we compare these two cases of tilapia and catfish, it is clear which error is more dangerous. In the tilapia case, many farmers might have spent a lot to purchase the new strain as per recommendation to replace the existing stock and also change the existing practices and facilities; whereas, in the case of catfish, farmers didn't need to change anything—they just followed their existing protocols. This means that there was no additional cost involved. Although both of these errors are unwanted and should be avoided, the case of tilapia (Type I error) is more dangerous than the case of catfish (Type II error).

6.6   Selection of statistical tools

The main purpose of teaching/learning statistics is to be able to choose the appropriate statistical tool(s) for a particular data set. Even if data have been generated perfectly, proper statistical analysis is needed to be able to make the right decisions. Misuse of statistical tools might result in wrong conclusions, thereby bad recommendations. Hence, even carefully collected data, which takes so much effort and time, can be useless. Researchers need to determine which distribution a particular data set follows: normal, binomial, Poisson, or free of any distribution. The fundamental principle is that, if the data sets are normally distributed sample means, SEs or SEs represent the population; but, if the data sets are distribution-free, then these parameters do not represent the population. Therefore, use of the mean values to represent and compare between/among them has no meaning. In such cases, nonparametric tests are used. These tests do not consider the actual figures and degree of deviations from central tendency but simply use ranks assigned to the data points. The following are the three main conditions suitable for the use of nonparametric tests:

- Data are far from normal or data do not follow any distribution pattern (normal, linear, binomial, or exponential), e.g. number of insects, bacterial count, disease incidence, salaries of staff, etc.
- Sample means, SDs, SEs, or variances do not represent the populations.
- Data are measured using ranks or other units, e.g. pH of water, grade point average or IQ of student, taste of fish, levels of social status, etc.

However, researchers should be clear about the assumptions and characteristics of nonparametric tests, which are as follows:

- Observations are independent of each other
- Scale of measurement is "rank"; therefore, ranking is done if data are not ranks (see Section 6.7.2)

Table 6.2 The names of tests their uses (detailed descriptions and methods are discussed in later sections).

Conditions or Purposes	Parametric Tests	Non-parametric Tests	
	Normal Distribution	Not-normal Distribution	Binomial Distribution
Compare a mean with standard value	One-sample *t*-test if $n < 30$, and *Z*-test if $n > 30$	Wilcoxon test	x^2- test or K–S test
Compare two means of unpaired data sets	*t*-test if $n < 30$, and *Z*-test if $n > 30$	Mann-Whitney test	x^2-test for large sample
Compare two means of paired data sets	Paired- sample *t*-test	Wilcoxon test	McNewman's test
Compare >2 means of unmatched data sets	One-way ANOVA and MRT	Kruskal-Wallis test	x^2-test
Compare >2 means of matched data sets	Multi-factor ANOVA and MRT	Friedman test	x^2-test
Find the relationship between two variables	Pearson's correlation	Spearman's correlation	Contingency Coefficients
Predict values of one variable from another	Simple linear or nonlinear regression	Spearman's correlation	Logistic regression
Find the relationship among several variables	Multiple regression (linear/nonlinear)	Kendall's coefficient of concordance	Multilogistic regression

- Have lower power than parametric tests; therefore, if parametric tests are not applicable, then these methods should be applied
- These methods are becoming more popular because occurrence of distribution-free data is quite common, especially in biological fields

Table 6.2 is a list summarizing all of the tests along with their uses. The list shows the names of nonparametric tests parallel to the parametric tests. As a researcher has to determine whether or not his/her data are in normal distribution before starting the statistical analysis, normality test should be the first step researchers take to decide whether to use parametric or nonparametric tests for a particular set of data.

6.7   Test of goodness-of-fit

How well a set of observations or the data match with a theoretical distribution is known as goodness-of-fit. Departure of observed data from the theoretical expected values occurs quite often. Using appropriate tools, these departures can be measured and tested to determine whether they are statistically still within the acceptable limit. A set of frequencies can be tested to determine whether it follows specific distribution, e.g. normal, binomial, and others. At the same time, data sets of two different samples can be tested to determine whether they follow the same pattern of frequency distribution. For this purpose, chi-square (x^2) and Kolmogorov–Smirnov (K–S) tests are used.

6.7.1 x^2-test

The x^2-test was first introduced by Karl Pearson in around 1900 and is usually used for discrete series following binomial and Poisson distributions. It tests a null hypothesis whether relative frequencies of occurrence of observed events follow any specific frequency distribution using the following formula:

$$x^2 = \Sigma[(O - E)^2/E]$$

Where,
   O is observed frequencies, and
   E is expected or theoretical frequencies.

The x^2-test is only for frequency data, not for percentages. If data are shown as percentages, then they must be converted to frequencies before using the x^2-test. Therefore, the test is useful only if the data are convertible to frequencies. Some uses of the x^2-test are given below.

a) **Test for normal distribution (normality test)**: The normality test is the gateway test which determines whether we should choose a parametric or non-parametric test. If the collected data are normally distributed, then parametric tests are used for hypothesis testing; but, if they are not normally distributed, then we have to either normalize them by using data transformation methods or use nonparametric tests instead. Normally, the x^2-test or K–S test (see Section 6.8.2) is used to determine whether the data set is normally distributed or not.

Table 6.3, for example, shows the size distribution of a sample taken from a fish population. It is necessary to test whether these data are normally distributed or whether the sample is a true representative of the population. The x^2-test can be used for this purpose. However, we must first establish a set of standardized (expected) frequencies, and then we can test for agreement or disagreement between observed and expected frequencies. To create the standardized frequencies from a given set of frequencies, the following formula is used:

$$f(x) = \exp[-0.5\{(X - m)/s\}^2]/[s \cdot \sqrt{(2p)}]$$

where,
   exp is exponential,
   X is data points,
   m is mean,
   p is pi (3.14), and
   s is SD.

Dotted lines in Figure 6.1 and Table 6.3 show the deviations in frequencies from the expected frequencies (solid lines). A x^2-test can determine whether these

Table 6.3 Size distribution of fish at harvest raised in a tank.

Size Group (g)	Observed Frequency	Expected Frequency	Standardized Frequency	$(O-E)^2/E$	
98	–	0.8	0.01014	–	
99	–	1.2	0.01506	–	
100	0	1.7	0.02146	1.72	
101	1	2.3	0.02933	0.77	
102	2	3.1	0.03846	0.38	
103	3	3.9	0.04836	0.20	
104	5	4.7	0.05834	0.02	
105	13	5.4	0.06750	10.70	
106	9	6.0	0.07491	1.51	
107	8	6.4	0.07974	0.41	
108	6	6.5	0.08142	0.04	
109	9	6.4	0.07974	1.08	
110	7	6.0	0.07491	0.17	
111	5	5.4	0.06750	0.03	
112	6	4.7	0.05834	0.38	
113	3	3.9	0.04836	0.20	
114	2	3.1	0.03846	0.38	
115	1	2.3	0.02933	0.77	
116	1	1.7	0.02146	0.30	
117	–	1.2	0.01506	–	
118	–	0.8	0.01014	–	
Total	1836	80	77	0.9680	19.05	$\rightarrow x^2$
Mean	108.0	5.33			16	$\rightarrow$ df
SD	4.9	3.46			26.3	$\rightarrow$ Table value
Var	24.0	11.95				

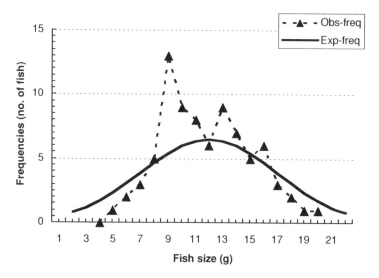

Figure 6.1 Fish size distribution with normal (standardized) curve.

Table 6.4 Number of fish counted for sex ratio test.

Sex	No. of Fish Counted or Observed (O)	No. of Fish Expected (E)	$(O-E)^2$	$(O-E)^2/E$
Male	38	30	64	2.13
Female	22	30	64	2.13
Total	60	60	128	4.27

deviations are still within the limits. Here, the x^2-test showed that probability is only 0.266.

Therefore, the researcher's decision will be "accept H_0," i.e. no difference between expected and observed frequencies. This means that, even though many frequencies seem to be away from the expected ones, the deviation can still be within the limit. Therefore, the sample is normally distributed, and it can serve as a representative for hypothesis testing using parametric tests.

b) **Test for binomial distribution:** Binomial distribution refers to a discrete series or nominal scale in which there are two (or sometimes more) mutually exclusive categories or classes, e.g. male and female, live and dead, etc. Normally, these data are in frequencies observed or measured, which can be tested against the expected ratio. For example, the ratio of male-to-female is 1:1 in normal population, if a student has a sample consisting of 38 males and 22 females. The x^2-test is used for this purpose in which the differences between the observed (O) and expected (E) numbers or frequencies are squared. The x^2 value is computed by summing the squared differences divided by their corresponding observed frequencies, as shown in Table 6.4.

Here,

$$x^2 = \Sigma[(O - E)^2/E] = 4.27$$
$$x^2_{1,0.05} = 3.84 \text{ (from Appendix A2)}.$$

The value 4.27 corresponds with the P value of 0.039 (<0.05). Therefore, the researcher will "reject H_0," which means that the observed ratio of male-to-female is significantly different from the normal ratio. This means that the number of males is significantly higher than the females in that population. This test can also be used for more than two categories.

c) **Heterogeneity test or contingency tables:** In a trial with formalin to protect fry from parasites, we found 77 alive and 33 dead with treatment compared with 55 alive and 47 dead without formalin (Table 6.5). We need to determine whether the formalin treatment had any effect or whether the survival/mortality was independent of the treatment.

In this case, expected frequencies are calculated. Then x^2 value in this case is 4.95, and the degree of freedom is 1.

Here, $x^2_{1,0.05} = 3.841$; therefore, "reject H_0" means that survival/mortality is affected by the formalin treatment. More specifically, results showed that

Table 6.5 Heterogeneity test on the effects of formalin treatment on fry survival.

	Observed			Expected		
	Untreated	Treated	Total	Untreated	Treated	Total
Live	55	77	132	$132 \times 102/212 = 68$	$132 \times 110/212 = 64$	132
Dead	47	33	80	$80 \times 102/212 = 42$	$80 \times 110/212 = 38$	80
Total	102	110	212	110	102	212

formalin increased fry survival from 53.9% ($55/102 \times 100$) to 70.0% ($77/110 \times 100$), i.e. 16.1% improvement.

6.7.2 One-sample K–S test

This method was developed by two Russian statisticians (Kolmogorov and Smirnov) to test the normality or the goodness-of-fit using cumulative frequencies based on the maximum difference in cumulative frequency (D_{max} or /d/). This means that it tests whether the highest deviation is still within the acceptable limit. Therefore, differences between observed cumulative frequencies and their respective expected frequencies are computed, as shown in Table 6.6, and the D_{max} is located and compared with the table value. These differences can be illustrated more clearly in graphical form, as shown in Figure 6.2.
Here,

$D_{max} = 11.60$
Number of observations $= 80$
Number of classes $= 17$
$D_{max\ 0.05,17,80} = 12.7$ (from statistical table, similar to given in Appendix A3)
K–S probability (P) $= 0.06$
Decision: Accept H_0, i.e. no difference between expected and observed frequencies.

Compared with the x^2-test, the one-sample K–S test is more precise and preferred, especially for the data set in a continuous series. However, the x^2-test is more appropriate for binomial distributions.

6.8 One- and two-sample tests

The most common and simplest form of hypothesis testing is the comparison between two means. Means are compared using the variability (Figure 6.3). If the sampled sets of data are normally distributed, as tested by K–S or as above, then parametric tests, e.g. Z- and t-tests, are used. If the data sets are not normally distributed, then nonparametric tests, e.g. Mann-Whitney and Wilcoxon's tests, are applied. Throughout this book, wherever possible, both parametric and nonparametric hypothesis testing have been described.

Table 6.6 Data on fish size distribution (from Table 6.3).

Size Group (g)	Observed Frequency	Expected Frequency	Standardized Frequency	Cumulative Frequencies		
				Obs-Cum	Exp-Cum	Difference
98	–	0.8	0.01014	–	–	–
99	–	1.2	0.01506	–	–	–
100	0	1.7	0.02146	0.0	1.7	−1.7
101	1	2.3	0.02933	1.0	4.1	−3.1
102	2	3.1	0.03846	3.0	7.1	−4.1
103	3	3.9	0.04836	6.0	11.0	−5.0
104	5	4.7	0.05834	11.0	15.7	−4.7
105	13	5.4	0.06750	24.0	21.1	2.9
106	9	6.0	0.07491	33.0	27.1	5.9
107	8	6.4	0.07974	41.0	33.4	7.6
108	6	6.5	0.08142	47.0	40.0	7.0
109	9	6.4	0.07974	56.0	46.3	9.7
110	7	6.0	0.07491	63.0	52.3	10.7
111	5	5.4	0.06750	68.0	57.7	10.3
112	6	4.7	0.05834	74.0	62.4	**11.6**
113	3	3.9	0.04836	77.0	66.3	10.7
114	2	3.1	0.03846	79.0	69.3	9.7
115	1	2.3	0.02933	80.0	71.7	8.3
116	1	1.7	0.02146	81.0	73.4	7.6
117	–	1.2	0.01506	–	–	–
118	–	0.8	0.01014	–	–	–
Total	1836	80			D_{max} value = 11.60	
Mean	108.0	5.33			No. of Classes = 34	
SD	4.9	3.46			D_{max} (tabulated) = 12.7	
Var	24.0	11.95				

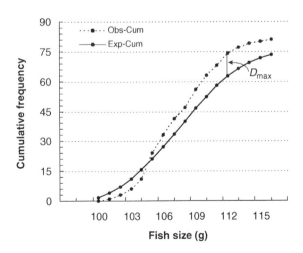

Figure 6.2 Observed and expected cumulative frequencies of fish size distribution.

83

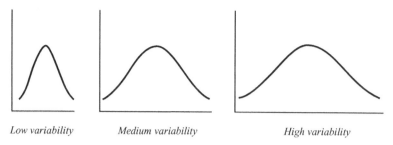

Low variability   Medium variability   High variability

Figure 6.3 Types of mean based on the variability.

6.8.1   Parametric tests: *t*- and *Z*-tests

A sample mean ($\bar{X}$) can be compared against its population mean (m) or against another sample mean. In order to compare with a standard population mean, the minimum sample size is considered 30. In this case, the test is called a *Z*-test. However, if the sample size is <30, then the Student's *t*-test is appropriate.

In other cases, a treatment mean is compared against zero in order to test the null hypothesis, which states, "there is no effect of treatment." Many researchers do not realize the value of this test. In many cases, experimental units can be limited in aquaculture. The trial can be designed for one sample test. For example, suppose a researcher has only four experimental units available. If he or she wants to compare two treatments and use two replications, there is a high chance of not detecting the significant difference between the two treatments due to lack of adequate replications. This means that there is a high possibility of committing Type II error. Instead of dividing four ponds among two treatments, another option is that all four ponds be used as replicates for a single treatment (described below).

6.8.1.1   One-sample *t*-test

A hypothesis that "tilapia does not grow in cold climate" was tested (Shrestha and Bhujel, 1999). We used four replicate cement ponds (Table 6.7) and conducted a trial using Nile tilapia (*Oreochromis niloticus*) in Nepal, where weather is normally considered cold and tilapia are not commonly accepted due to fear of low temperature and poor growth. The mean daily weight gain from the four replicates (1.0 g) was compared against zero (0).
Here,

H_0: Tilapia does not grow in cold climate (0 g·fish^{-1}·day^{-1})
H_A: Tilapia can grow even in cold climate (>0 g·fish^{-1}·day^{-1})
Mean weight gain ($\bar{X}$) = 1.0 g·fish^{-1}·day^{-1} and SD (s) = 0.1 g·fish^{-1}·day^{-1}
Hypothetical mean (m) = 0 g·fish^{-1}·day^{-1}(no growth) *t*-statistic = $(x - m)/$
$(s/\sqrt{n}) = (1.0 - 0.0)/(0.1/\sqrt{4}) = 20.0$
From the *t*-table, $t_{0.05, 3} = 3.182$, which means $P < 0.05$
Decision: Reject H_0

This means the daily weight of tilapia was significantly higher than zero, or tilapia grew in cold climate.

Table 6.7 Results of a trial on tilapia that used a one-sample *t*-test.

	1	2	3	4	Mean ± SE
		Replicate Ponds			
Initial weight (g)	44.1	47.2	33.3	33.3	39.6 ± 3.7
Final weight (g)	148.1	152.1	129.7	150.7	145.2 ± 5.2
Weight gain (g·fish^{-1}·day^{-1})	1.0	1.0	0.9	1.1	1.0 ± 0.0
Survival (%)	94.4	77.8	94.4	94.4	90.0 ± 4.2

6.8.1.2   *Z*-test

This test compares a population mean (m) with a sample mean ($\bar{X}$) using population SD (s). For example, productivity of pond culture (t·ha^{-1}) obtained from a survey of 35 farms in a district is compared with the national average recorded from the country's census. The method of analysis is given below, together with Student's *t*-test.

6.8.1.3   Student's *t*-test for two-sample means

The Student's *t*-test was developed by W. S. Gossett (1876–1937), who considered and nicknamed himself "Student" because he always felt that he was in the learning phase. The test is used for the comparison of any two means using sample SDs, especially when the number of observations is less than 30. There are two types of Student's *t*-test:

- independent samples
- paired/matched samples

When performing a *t*-test, statistics is calculated and compared with the tabulated value for "*P*" using pooled variance. If the sample size is different, pooled variance is calculated by taking the weighted average of the variance as shown below.

$$t_{n_1+n_2-2} = (\bar{Y}_1 - \bar{Y}_2)/\text{pooled variance}$$
$$\text{Population pooled variance} = s_1^2/n_1 + s_2^2/n_2$$
$$\text{Sample pooled variance} = S_1^2/n_1 + S_2^2/n_2$$
$$df = n_1 + n_2 - 2$$

For an example, suppose a fish nursing trial was conducted using eight tanks for 1 month to compare two types of feed in the growth of fish. The final mean weights were obtained as follows:

Homemade feed (G1) = 50.3 ± 10.1 g/fish (Mean ± 1 SD)

Commercial feed (G2) = 69.1 ± 9.2 g/fish

Difference in means = 69.1 − 50.3 = 18.8 g, i.e. 37% bigger than the fish of G1, if it is statistically proved this difference has biological meaning. But when compared using a *t*-test, this difference is not statistically significant, as shown below.

$$t\text{-statistics} = (\bar{Y}_1 - \bar{Y}_2)/[(S_1^2/n_1) + (S_2^2/n_2)]$$
$$= (69.1 - 50.3)/(10.1^2 + 9.2^2) = 0.1007$$
$$t_{0.05,6} = 2.447 \text{ (from Appendix A4)}$$

The result is "accept H_0", which means the final weights of fish fed with two types of feed were not significantly different ($P > 0.05$). The difference of 18.8 g is not statistically significant because of high SD. Therefore, either feed can be recommended.

Another example is a survey on aquaculture productivity in two districts (Districts A and B), which distinguishes between the two tests (Table 6.8). It was conducted using a standard questionnaire. At the same time, the national average for fish productivity 4.2 t·ha^{-1}·year^{-1} with the standard deviation (SD$_P$) 0.6 t·ha^{-1}·year^{-1} was obtained from the statistical bureau. In this case, comparison can be made between Districts A and B using the *t*-test and between District A or B against national average using the Z-test, as described below.

Both data sets are presented in graphical form (Figure 6.4), in which they look normally distributed. However, in order to confirm this normal distribution, they can be tested using the K–S test. First, they need to be grouped as shown in Table 6.9.

$$D_{\max\ 0.05,\ 11,\ 35} = 8 \quad \text{and} \quad D_{\max\ 0.05,\ 12,\ 35} = 8$$

The maximum differences in cumulative frequency in both distributions are lower than this value. Therefore, both data sets are in normal distribution, so we can proceed with comparisons.

a) Comparison between the national average and District A:

$$Z = (4.20 - 3.05)/(\sqrt{0.95/35 + 0.6}) = 1.45$$

b) Comparison between the national average and District B:

$$Z = (5.24 - 4.20)/(0.6) = 1.7$$

c) Comparison between Districts A and B:

$$t = (5.24 - 3.05)/[\sqrt{(0.90 + 0.74)/35}] = 10.09$$

Table 6.10 presents the results of the Student's *t*-test and the Z-test. The Z-test was performed by using the variance of national average only. But to compare between the two districts (i.e. A and B), their pooled variance was used.
Notes on writing results: Statistical analysis only assists in writing results and making conclusions and recommendations or in other words, making decisions.

Table 6.8 Data collected from the survey of fish production in Districts A and B, and frequency distributions.

Farm No.	District A	District B
1	4.1	6.5
2	3.2	5.5
3	2.1	4.5
4	3.0	4.4
5	3.6	4.5
6	3.6	4.6
7	4.3	5.6
8	4.3	5.5
9	4.4	5.4
10	4.9	5.5
11	3.6	5.4
12	3.1	4.3
13	1.2	4.1
14	2.3	4.6
15	3.2	5.6
16	2.3	5.6
17	3.3	5.4
18	2.3	5.0
19	2.1	6.8
20	5.1	6.5
21	1.3	5.4
22	2.1	5.4
23	2.6	5.2
24	2.2	4.8
25	2.3	4.5
26	3.2	4.4
27	4.3	3.4
28	3.2	3.9
29	3.3	4.8
30	3.1	4.9
31	2.3	6.9
32	3.3	6.8
33	2.3	5.4
34	2.2	6.5
35	3.2	5.8
Mean	3.05	5.24
SD	0.95	0.86
Variance	0.90	0.74
Skew	0.236	0.218
Kurtosis	−0.354	−0.315

However, interpretation and presentation of the results largely depend on the degree of skills possessed by individual researchers. While writing results, many researchers often miss important information obtained from statistical analysis, thus resulting in inadequate explanation, whereas others may misuse them by saying higher or lower even though they are not statistically different. The following points should be helpful for using the results of the survey:

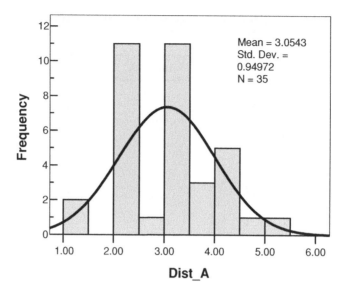

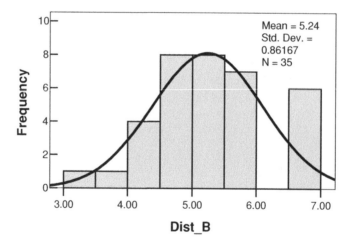

Figure 6.4 Frequency distribution of fish farms based on productivity (t·ha^{-1}·year^{-1}).

- Beginners may write, "Two districts have significantly different ($P < 0.05$) fish productivity" or "they differ significantly ($P < 0.05$) in fish productivity." These statements are neutral as these do not specify which district has higher productivity and by how much or what percentage. Therefore, these sentences do not adequately state or communicate the result.
- Results might also be written as, "District B has significantly higher ($P < 0.05$) fish productivity compared with District A." This specifies which district has higher productivity; however, it still does not compare and quantify the difference in terms of the amount or percentage.
- A better way of stating the result is, "District B has 2.19 t·ha^{-1}·year^{-1} (5.24 − 3.05) or 72% (2.19/3.05 × 100) higher fish productivity than District A."

Table 6.9 Grouping of fish farms based on their productivity for K–S test.

District A							
	Class	Obs-Freq	Exp-Freq	St-Ffreq	Obs-Cum	Exp-Cum	D_{max}
	1.0	0	1.2	0.0353	0.0	1.2	−1.2
	1.4	2	1.9	0.0537	2.0	3.1	−1.1
	1.8	0	2.6	0.0751	2.0	5.7	−3.7
	2.2	5	3.4	0.0966	7.0	9.1	−2.1
	2.6	7	4.0	0.1143	14.0	13.1	0.9
	3.0	1	4.3	0.1243	15.0	17.5	−2.5
	3.4	10	4.3	0.1243	25.0	21.8	3.2
	3.8	3	4.0	0.1143	28.0	25.8	2.2
	4.2	1	3.4	0.0966	29.0	29.2	−0.2
	4.6	4	2.6	0.0751	33.0	31.8	1.2
	5.0	1	1.9	0.0537	34.0	33.7	0.3
	5.4	1	1.2	0.0353	35.0	35.0	0.0
Total	38.4	35	35.0	0.99858			

District B							
	Class	Obs-Freq	Exp-Freq	St-Freq	Obs-Cum	Exp-Cum	D_{max}
	3.0	0	1.4	0.039	0.0	1.4	−1.4
	3.4	1	2.2	0.062	1.0	3.5	−2.5
	3.8	0	3.1	0.087	1.0	6.6	−5.6
	4.2	2	3.9	0.112	3.0	10.5	−7.5
	4.6	8	4.6	0.130	11.0	15.1	−4.1
	5.0	4	4.8	0.137	15.0	19.9	−4.9
	5.4	7	4.6	0.130	22.0	24.4	−2.4
	5.8	7	3.9	0.112	29.0	28.4	0.6
	6.2	0	3.1	0.087	29.0	31.4	−2.4
	6.6	3	2.2	0.062	32.0	33.6	−1.6
	7.0	3	1.4	0.039	35.0	35.0	0.0
Total	55.0	35	35.0	0.999			

- The best way of stating the result is, "Fish productivity in District B is 5.24 t·ha^{-1}·year^{-1}, which is 72% higher ($P < 0.05$) than in District A, wherein productivity is only 3.05 t·ha^{-1}·year^{-1}. This statement provides the productivity of both districts, specifies which district has higher productivity, and also points out by what percentage that amount is higher.
- Similarly, comparison of Districts A and B against the national average should not be forgotten. Based on the results, it can be concluded that "productivity of Districts A and B differs significantly ($P < 0.10$; as it is a social survey research, 10% significance level can be used instead of 5%) compared with the national average." District A has 27.4%[(4.2 − 3.05)/4.2 × 100] lower, whereas District B has 24.8% [(5.24 − 4.2)/4.2 × 100] higher productivity than the national average.

Table 6.10 Results of Z- and t-test using Microsoft® Excel.

Z-test	District A	National Average
Mean	3.05	4.20
Known Variance	0.9	0.36
Observations	35	1
Z	1.845	
$P(Z \leq z)$ two-tail	0.065	

Z-test:	District B	National Average
Mean	5.24	4.20
Known Variance	0.74	0.36
Observations	35	1
Z	1.68	
$P(Z \leq z)$ two-tail	0.092	

t-test:	District A	District B
Mean	3.054	5.24
Variance	0.902	0.74
Observations	35	35
Df	67	
t-statistics	10.084	
$P(T \leq t)$ two-tail	4.57E-15	

6.8.1.4 Paired-sample *t*-test

A paired-sample *t*-test is used to compare sample means of paired data sets generated from the same or related subjects over time or in differing circumstances. In order to be paired, the samples of all the treatments should be drawn simultaneously (Figure 6.5). For example, temperature and dissolved oxygen (DO) recorded at certain intervals of time, such as every morning/afternoon, weekly, monthly, and so on. Several examples of paired samples can be found in research, but most of the time it is ignored. The *t*-statistics are computed as:

$$t = \text{mean difference between pairs/SE of mean difference} = d/s\sqrt{n}$$

Where
  n = number of pairs

The data shown in Figure 6.5 can be compared using a paired-sample *t*-test. Another aspect of these data is to determine whether they have any correlation (see Section 8.4 for details). In order to make these clear, it can be described as:

1. Data sets A and B are most likely to have both significant difference and correlation.
2. Data sets A and C are most likely to have difference but no correlation.
3. Data sets B and C are most likely to have neither correlation nor difference.

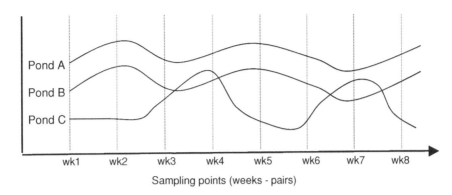

Figure 6.5 Weekly temperatures (or DO, pH levels) of three ponds measured at the same time on the same day of each week.

There is a chance of having significant difference between data sets A and B when analyzing them by using a paired-sample *t*-test on the basis that all of the data points of Pond A show higher compared with the corresponding data points for Pond B. If these two ponds were compared by using an independent sample *t*-test, results would show no difference between them because it would mess up the data of whole Pond A while comparing with the pooled data of Pond B, rather than comparing for each corresponding point. When pooled, lowest points of Pond A are compared with the highest points of Pond B, which do not differ. The overlap paired *t*-test would not pool them; instead, it measures the differences and compares all points separately. Therefore, it is appropriate.

For example (data are presented in Table 6.11), a breeding trial with tilapia was conducted over a period of 13 weeks using normal and prestunted broodstock to compare their reproductive performance. Broodfish were fed at 3% biomass. Seed (number of eggs or post-larvae) was harvested on a weekly basis. Considering the weekly harvesting as pairs, they can be analyzed by using a pair-sample *t*-test.

SD can be computed from variance, e.g. SD for the mean egg output for stunted tilapia $= \sqrt{902847.6} = 950$.

The two-tail probability (**0.002048) shows that $P < 0.01$; therefore, the difference in seed output is highly significant. Therefore, the result can be written as, "Average seed output from stunted tilapia $(1,345 \pm 950)$ is about 275% $(1,345/490 \times 100)$ higher $(P < 0.01)$ than that produced from normal tilapia in this trial."

Table 6.11 Seed output (no. of eggs or post-larvae) per week from normal and stunted Nile tilapia broodfish.

Weeks	1	2	3	4	5	6	7	8	9	10	11	12	13
Normal	44	339	47	475	373	1,026	109	572	874	351	109	1,769	283
Stunted	393	413	297	800	1,509	2,183	1,262	0	3,108	2,109	1,094	2,293	2,027

Table 6.12 Results of paired- sample *t*-test performed with Microsoft® Excel using the data analysis function.

t-test: paired two-sample for means	Normal Tilapia	Stunted Tilapia
Mean	490	1345
Variance	238430.4	902847.6
Observations	13	13
Pearson correlation	0.561859	
Hypothesized mean difference	0	
Df	12	
t-statistic	−3.916572	
$P(T \leq t)$ one-tail	0.001024	
t-critical one-tail	1.782288	
$P(T \leq t)$ two-tail **	**0.002048**	
t-critical two-tail	2.178813	

6.8.2   Nonparametric tests: Mann-Whitney and Wilcoxon's tests

As described in earlier sections, nonparametric tests are alternative tests to be used only when data are not normally distributed. These tests use ranks for analysis rather than the actual values. Therefore, ranking is performed on the original values before analyzing the data. For the purpose of describing this method, an example is used (Table 6.13) in which two carps, Rohu and Catla, are compared based on weight (g) after growing together for 1 year.

Mean weights of Rohu (g): 292, 287, 267, 282, 279, 286

Mean weights of Catla (g): 276, 272, 261, 266, 263, 255

Sorting data by either ascending or descending order is the prerequisite for ranking. Ranking is done for all of the data or observations of all the groups together, if data are not paired. For the data set arranged in ascending order, rank 1 is

Table 6.13 Simple ranking of unpaired data on the weight of carps.

Weight (g)		Ranks	
Catla	**Rohu**	**Catla**	**Rohu**
276	292	6	1
272	287	7	2
266	286	9	3
263	282	10	4
261	279	11	5
255	267	12	8

Table 6.14 Tied ranking of fish weight data with repeated values.

Weight (g)		Ranks	
Catla	Rohu	Catla	Rohu
237	245	7.5	11
235	248	3.5	13
238	237	7.5	6
245	235	11	3.5
242	251	9	14
236	245	5	11
225	255	2	15
222	258	1	16

given to the lowest value, rank 2 to the second lowest, and so on. Opposite will be the case if data are arranged in descending order.

If some of the observations are repeated, then they are called tied, and the repeated observations get the average of their ranks. For example, in Table 6.14, there are two values of 235, which should be ranked 3 and 4; instead, they are averaged and their rank is then 3.5. Similarly, three values of 245 with ranks 10, 11, and 12 are assigned their average rank, i.e. 11.

6.8.2.1   Mann-Whitney test (U-test)

Although it was originally developed by Wilcoxon, Mann and Whitney further developed the method; therefore, it is commonly known as the Mann-Whitney test (Zar 1996). This method is used to compare two independent samples, similar to the Student's t-test for two independent samples in parametric testing. The Mann-Whitney statistic (U) is calculated as:

$$U = (n_1 \times n_2) + [n_1 \times (n_1 + 1)/2] - R_1$$

Where,
n_1 is the number of samples in the first group,
n_2 is the number of samples in the second group, and
R_1 is the sum of the ranks of the first group.
Here, it is assumed that $n_1 > n_2$, but if $n_2 > n_1$, then the equation should be:

$$U = (n_1 \times n_2) + [n_2 \times (n_2 + 1)/2] - R_2$$

Where,
R_2 is the sum of the ranks of the second group.
An example of nominal data for the comparison between Rohu and Catla (Table 6.15) is given to demonstrate this method.
H_0: There is no difference between the total weights of Rohu and Catla.

Table 6.15 Ranking of weights of Rohu and Catla.

Weight (g)		Ranks	
Catla	**Rohu**	**Catla**	**Rohu**
237	245	8	10.5
233	248	3.5	12
237	237	8	8
233	235	3.5	5
222	251	1	13
236	245	6	10.5
225	255	2	14
232	**245**		
	Rank totals	**32**	**73**

Here,

$$n_1 = 7$$
$$n_2 = 7$$
$$U = \{n_1 \times n_2 + [n_1 \times (n_1 + 1)/2]\} - R_1$$
$$U = [7 \times 7 + (7 \times 8)/2] - 32$$
$$= 45$$

$U_{0.05,7,7} = 41$ (from U table, Appendix A5)
Therefore, "reject H_0," which means the mean weight of Rohu is significantly higher than that of Catla ($P < 0.5$). This method is specifically used for ordinal data as shown in Table 6.16, where the H_0 claims that there is no difference between the tastes of farmed and wild catfish.
Here,

$$n_1 = 9$$
$$n_2 = 8$$
$$U = [9 \times 8 + (9 \times 10)/2 - 69.5$$
$$= 47.5$$

$U_{0.05,8,9} = 57$ (from U table, Appendix A5)

Therefore, "accept H_0," which means there is no difference between scores given for the taste of farm-reared and wild catfish; in other words, there is no difference in taste.

Table 6.16 Ranking of taste of wild and farm-reared catfish collected from wild and farmed.

Scores (Data)		Ranks	
Farmed	**Wild**	**Farmed**	**Wild**
A	A	14.5	14.5
B+	A	12	14.5
B+	A	12	14.5
B	B+	8	12
B	B	8	8
B	B	8	8
C+	C+	4	4
C	C+	1.5	4
C			1.5
Rank Totals		69.5	79.5

6.8.2.2   Wilcoxon's test for paired samples

This test is also called "Rank Sum," "Matched pair," and "Signed Rank" test; it is analogous to the paired *t*-test but with low power, which is true for all nonparametric tests.

An example is the comparison of the growth of two types of tilapia after rearing together in 10 tanks. Their initial weights were similar. After rearing for 120 days with 3% body weight of feeding, final mean weights were obtained, as shown in Table 6.17.
Here,

H_0 is that there is no difference in mean final weights (g) between two types of tilapia.
Here,

$$T_+ = 7 + 4.5 + 2 + 4.5 + 7 + 9.5 + 7 + 9.5 = 51.0$$
$$T_- = 1 + 3 = 4$$

Table 6.17 Comparison of growth between Nile and Red tilapias.

Tank No.	Nile Tilapia	Red Tilapia	Difference (d)	Assigned Rank of d
1	121	116	5	7
2	119	120	−1	−1
3	119	115	4	4.5
4	125	123	2	2
5	117	113	4	4.5
6	121	124	−3	−3
7	123	118	5	7
8	126	120	6	9.5
9	127	122	5	7
10	119	113	6	9.5

$T_{0.05,10} = 8$ (from Appendix A6) and $P < 0.05$. Therefore, "reject H_0," which means the final mean weight of Nile tilapia is higher than that of Red tilapia.

6.9   Questions

Q1. Why is hypothesis testing important?
Q2. Why are errors important and how can you avoid them?
Q3. Why is the selection of appropriate statistical tools important?
Q4. What are the differences between statistical and biological significance?
Q5. In what ways are nonparametric tests useful?

6.10   Practical exercises

Ex. 1. Table 6.18 shows the results of five pairs of cross-breeding using Nile tilapia (*Oreochromis niloticus*) and Java tilapia (*O. mossambicus*). The expected ratio of progeny is 3:1 (black-to-red). Test whether each ratio and the combined ratios are similar to the expected one.

Ex. 2. In your study area, there are 3,900 men and 4,000 women. Test whether the ratio of women-to-men is significantly higher or not.

Ex. 3. A feed company claims that a newly formulated commercial feed can reduce feed conversion ratio (FCR) in catfish by almost 25%. An average of 1.8 is commonly obtained with the existing feed. Believing this claim, 10 farmers used the new feed. Collected FCR data were as follows: 1.5, 1.6, 1.8, 1.2, 1.3, 1.2, 1.2, 1.3, 1.2, and 1.3. Test whether the feed company's claim is supported by these data and statistical analysis.

Ex. 4. DO levels ($mg \cdot L^{-1}$) were measured from two ponds at early morning (6 a.m.) over a period of 9 weeks (Table 6.19). Pond A is located close to a few trees that give some shade, but Pond B is in a more open area. Test whether these two ponds differ in terms of DO level by using the data provided.

Ex. 5. Two strains of Nile tilapia (local and improved) were compared for their reproductive performance. Four groups of similar sizes were used as replicates. The average numbers of eggs produced per spawning over the 3-month period are shown in Table 6.20. Test whether these two types of tilapia broodstock differ significantly in reproductive performance.

Table 6.18 Number of fry produced by five pairs of crosses.

Fish Pairs	1	2	3	4	5
Black fry	350	633	456	45	412
Red fry	133	160	145	23	99

Table 6.19 Weekly DO levels of two ponds.

| | \multicolumn{9}{c}{DO at 6 a.m.} | | | | | | | | |
	Wk 1	Wk 2	Wk 3	Wk 4	Wk 5	Wk 6	Wk 7	Wk 8	Wk 9
Pond A	0.6	0.6	0.5	0.1	0.2	0.1	0.1	0.1	0.0
Pond B	1.8	1.8	1.6	1.5	1.3	1.2	1.0	0.9	0.7

Table 6.20 Average seed output of local and an improved strain of tilapia.

Strain	Group 1	Group 2	Group 3	Group 4
Local	1,405	1,576	1,537	1,345
Improved	1,612	1,646	1,599	1,688

Chapter 7

Experimental designs and analysis of variance

7.1 Background

When a hypothesis is tested by comparing variances after partitioning, the method is called analysis of variance (ANOVA). More specifically, the effect of any factor is considered significant if the variance of a treatment is higher than the variance among the replicates. Various experimental designs are used to separate these variances. Application and explanation of these designs are presented in this chapter.

7.2 Completely randomized design

Completely randomized design (CRD) is the basic experimental design that is used to study the effects of one factor, i.e. treatment or fixed factor, keeping others constant; therefore, it is often called a single-factor experiment. The selected factor (e.g. feed) is varied to form at least three types of different treatments, e.g. commercial feed, farm-made feed with fish meal, farm-made feed without fish meal, etc. The changes in response variable(s) caused by the different factors are observed/measured, e.g. survival, growth, net fish yield, etc. For CRD, all of the experimental units should be uniform, and the types of selected factors (treatments) are randomly assigned to the experimental units. Allocation of the treatments and replications can be done by using a lottery system, random numbers/table, or any other method. Before randomizing, we need to determine the required total number of experimental units (n). If there are "t" different treatments of a single factor and the treatments are replicated "r" times, then:

Total experimental units $(n) = t \times r$

For example, if we want to compare 4 feed types with 5 replications, we will need $4 \times 5 = 20$ uniform tanks (Figure 7.1), or any other experimental units. Treatments are coded as T1, T2, T3, and T4, and similarly replicates are R1, R2, R3, R4, and R5. The combination of treatments and replicates is coded as T1R1, T1R2, ..., T4R5, as shown in Table 7.1. All of the tanks are assigned with a number (1–24) for randomization. With the lottery method, all of these treatment–replication combinations are written on 20 uniform pieces of paper and kept in one bag. Similarly, tank numbers are written on another set of 20 uniform pieces of paper and kept in another bag. Assigning the tank for each treatment–replication combination can be done by randomly picking the treatment–replications (with one hand) and the tank numbers (with the other hand) from both bags simultaneously. An example of treatment randomization is shown in Figure 7.1.

Once randomization is complete, the researcher can start the experiment and collect required data depending on the specific objectives of the research. Some examples of response variables include:

- fish survival (record any dead fish observed over the period of the experiment)
- fish growth and yield (batch weights, individual weights of sampled fish from each tank or experimental unit)
- differences in body composition; samples need to be taken from muscles, bones, blood, etc.

Table 7.1 Treatment combinations or experimental design for CRD.

| Replication | Treatments | | | |
	T1	T2	T3	T4
R1	T1R1	T2R1	T3R1	T4R1
R2	T1R2	T2R2	T3R2	T4R2
R3	T1R3	T2R3	T3R3	T4R3
R4	T1R4	T2R4	T3R4	T4R4
R5	T1R5	T2R5	T3R5	T4R5

Tank 1	Tank 2	Tank 3	Tank 4
T3R2	T2R3	T4R5	T1R3
Tank 5	Tank 6	Tank 7	Tank 8
T4R1	T4R3	T3R5	T3R1
Tank 9	Tank 10	Tank 11	Tank 12
T1R2	T1R1	T1R5	T2R2
Tank 13	Tank 14	Tank 15	Tank 16
T2R1	T4R4	T4R2	T2R5
Tank 17	Tank 18	Tank 19	Tank 20
T3R3	T2R4	T1R4	T3R4

Figure 7.1 Complete randomization of treatments in all of the experimental units.

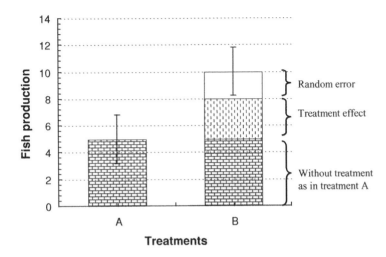

Figure 7.2 Separation of effects of treatment and experimental or random error.

There are several factors, especially environmental and water chemistry, e.g. temperature, DO, pH, alkalinity, ammonia levels, nitrite levels, etc., that can't be controlled, but their levels may vary due to the treatments or other external factors. However, these factors have direct impacts on the main response variables, such as survival, growth/yield, and the body composition of fish. Therefore, these factors should also be measured or recorded. They can be used as covariates (see Section 9.2), which will assist in interpreting the results.

The following equation represents the mathematical model for CRD:

$$Y_{i,j} = m + T_i + R_i$$

Where,

$Y_{i,j}$ is any observation for which $X_1 = i$
i is the level of the factor
j is the replication within the level of the factor
m is the general location parameter
T_i is the effect of having treatment level i
R_i is the random error at treatment level i

Basically, in order for the treatment effect to be statistically significant, the increment in treatment B (Figure 7.2) due to treatment needs to be higher than the experimental or random error.

7.2.1   Parametric test

Normally distributed data are analyzed by using a simple one-way ANOVA. The steps are as follows:

1. Group the data by treatments and calculate the treatment totals (T), grand total (G), grand mean, and coefficient of variation (CV).

Table 7.2 ANOVA table for CRD.

Source of Variation	SS	df	Mean Square	F	P
Treatments (factor)	SST	$t - 1$	$\dfrac{SST}{t - 1} = MST$	$\dfrac{MST}{MSE}$	$P < 0.05^*$ $P < 0.01^{**}$
Residual error	SSE	Total df − trts df	$\dfrac{SSE}{t(r - 1)} = MSE$		$P > 0.05^{ns}$
Total	SS	$tr - 1$			

2. Using the number of treatments (t) and the number of replications (r), determine the df for each source of variation.
3. Construct an outline/table of ANOVA as shown in Table 7.2.
4. Using X_i to represent the measurement of the ith plot, T_i as the total of the ith treatment, and n as the total number of experimental plots [i.e. $n = rt$], calculate the correction factor (CF) and the various sums of square (SS).
5. Calculate the mean square (MS) for each source of variation by dividing SS by their corresponding df.
6. Calculate the F-value (R.A. Fisher) for testing significance of the treatment difference, i.e. mean square of treatment divided by the mean square error $(F = MST/MSE)$.
7. Enter all of the computed values in the ANOVA table.
8. Obtain the tabular F-values using:
   $f_1 = $ treatment $df = (t - 1)$
   $f_2 = $ error $df = t(r - 1)$
   and compare, as shown in Table 7.3.

CRD is very important because it has a high proportion of df; therefore, it is suitable for smaller trials with fewer experimental units. ANOVA is done before performing multiple range tests, as it is stronger than multiple range tests to see the effects of a factor, but ANOVA does not compare means, nor does it locate differences. On the other hand, if experimental units are not homogenous,

Table 7.3 Basis of conclusions to be made.

t- or f-Values	P-Value	Significance	Signs Used	Inference
Calculated value < tabulated value for 0.05	$P > 0.05$	Nonsignificant	ns	Accept H_0
Calculated value > tabulated value for 0.05	$P < 0.05$	Significant	$*$	Reject H_0
Calculated value > tabulated value for 0.01	$P < 0.01$	Highly significant	$**$	Reject H_0

Table 7.4 Data collected from a trial to compare four feeds.

Replicates	Feed 1	Feed 2	Feed 3	Feed 4	Total
1	86	88	105	94	
2	83	87	104	91	
3	91	94	102	89	
4	84	86	99	92	
5	87	89	–	97	
Total	431	444	410	463	1,748
n	5	5	4	5	19
Mean	86.2	88.8	102.6	92.5	

there will be an increased experimental error. In such a case, other designs must be applied as described in the following sections.

An example of analysis for CRD is presented here. Four iso-protein commercial catfish pellets differing in lipid levels were compared by stocking 50 fish in each of the cages installed in a single large pond. Data shown in Table 7.4 are the mean final weights (g) of fish (fish from a cage in replicate 5 assigned for Feed 3 escaped during handling). Now we can test whether these four feeds (treatment) show any differences in terms of fish growth.

$$\text{Here, } H_0 = m_1 = m_2 = m_3 = m_4$$

Step 1: Calculate sum squares.
Correction factor (C) = (Grand total)^2/n = $(1,748)^2$/19 = 160,816
Total SS = $(86)^2 + (88)^2 + \cdots + (97)^2$ – C
         = 161,609 – 160,816 = 793
Treatment SS = Σ [(Treatment total)^2/n] – C
             = $(431)^2$/5 + $(444)^2$/5 + $(410)^2$/4 + $(463)^2$/5 – 160,816
             = 665
Error SS = Total SS – Treatment SS = 793 – 665 = 128

Step 2: Prepare an ANOVA table as shown in Table 7.5.
From the F-Distribution table shown in Appendix A7, $F_{3,15,0.01}$ = 5.42

Table 7.5 ANOVA table of fish growth comparison based on data from Table 7.4.

Sources of Variation	(SS)	df	Mean SS = SS/df	F = MST/MSE	P	Significance
Total	793	18				
Treatment	665	3	222	26	<0.01	**
Error or residual	128	15	9			

Note: Numerator df = 3; Denominator df = 15.

Reject H_0 ($P < 0.01$) which means that the treatment (feed) has highly significant ($P < 0.01$) effect on fish growth. However, ANOVA does not show which feed is the best among all or which ones are better than others. In order to compare among the means for each feed, multiple comparisons are necessary. On the other hand, if an ANOVA shows that there is no effect of the factor (feed), then there is no need to compare among means; multiple comparisons are unnecessary.

Multiple Comparison Tools: Multiple comparisons are used to compare all possible combinations of means simultaneously and are often called pairwise comparisons, based on which ranking among the means is possible. There are several methods for multiple comparisons. The most commonly used methods are least significant difference (LSD), Newman-Keuls test, Duncan's multiple range test (DMRT), Tukey's honestly significant difference (HSD) Test, and Scheffe test (Zar 1996). LSD is not suggested for multiple comparisons, although it was used by many researchers. Scheffe test is best for multiple contrasts. Therefore, DMRT and Tukey's HSD are most popular for multiple comparisons. Tukey's HSD is better if data are less normal. Nevertheless, results of these tests are more reliable and robust when the data sets are more normal and equal in sample sizes. If there is doubt that data are not in normal distribution, then nonparametric multiple comparisons are suggested (discussed in later sections).

The basic principle in multiple comparisons or pairwise comparisons is that a common value for difference using pooled variance is calculated. Different methods have been devised for specific purposes and situations. Most statistical software include these options and can be easily used. With a view to describing the method, comparisons among means have been made. First of all, a value for comparison has to be computed from the pooled SE multiplied by the critical value for the given significance level. The common value of 4.0 has been obtained for the comparison between two means at time as follows:

$$
\begin{aligned}
SE_{\bar{X}_1 - \bar{X}_2} &= \sqrt{S^2(1/N_1 + 1/N_2)} \\
&= \sqrt{9(1/5 + 1/5)} \\
&= 1.9\ g
\end{aligned}
$$

$t_{0.05,\,15\,df} = 2.131$, 95% CI $= 1.9 \times 2.131 = 4.0\ g$

Results can be presented either in tabular form (Table 7.6) or in graphical form (Figure 7.3). Just one form would be enough; however, for the purpose of describing, both are shown here.

As a rule of thumb, if the difference between any two means is higher than 4.0, then those means are considered significantly ($P < 0.05$) different. However,

Table 7.6 Table showing comparisons among the means.

	Feed 1	Feed 2	Feed 4	Feed 3
Means	86.2	88.8	92.5	102.6

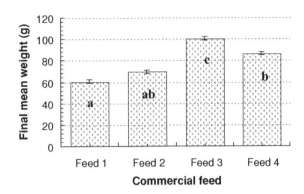

Figure 7.3 Final graphical presentation of result using error bars for variability.

the researcher should specify which means are higher or lower than which ones. Using this example, the following conclusions can be drawn:

- Feed 3 resulted in the highest ($P < 0.05$) growth (mean weight) of fish.
- Feed 4 resulted in the higher ($P < 0.05$) fish growth compared with Feed 1.
- Feed 4 and Feed 2 didn't differ ($P > 0.05$) in terms of fish growth.
- Feed 1 and Feed 2 didn't differ ($P > 0.05$) in terms of fish growth.

While making recommendation, it should be mentioned that "based on the results of the trial, Feed 3 should be recommended for highest growth. However, economic value of using this feed or its price as compared with others needs to be considered." As a basic principle, the additional cost of using this feed should be less than the economic value of the added yield in the fish production over others. This is generally neglected by most researchers, and they may directly recommend the feed that gives the highest growth and biological significance. Economic analysis is beyond the scope of this book, so researchers who need further explanation should refer to relevant books or consult relevant economists.

7.2.2 Nonparametric test: Kruskal-Wallis test (*H*-test)

Nonparametric tests are similar to parametric tests for ANOVA; but, they use ranks rather than the original data for analysis. Therefore, they are also called "ANOVA by ranks." When the samples are not from normally distributed data or the variances are heterogeneous, ranks are assigned to the observation for analysis. The method of ranking is done in the same way as described in Section 6.8.2. As in parametric tests, the Kruskal-Wallis test only determines whether there is an effect by a factor, but it doesn't compare among the means. A nonparametric method has also been developed for the purpose of multiple comparisons.

To describe these steps, an example has been taken from Zar (1996) in which pH data from eight samples in four ponds (Table 7.7) are collected by a limnologist with the aim of comparing four different ponds in terms of pH. Whether there are any pH differences among pH values among ponds is tested.

Table 7.7 Data on pH values of water sampled from four ponds.

Rep.	Pond 1	Pond 2	Pond 3	Pond 4
1	7.68 (1)	7.71 (6)	7.74 (13.5)	7.71 (6)
2	7.69 (2)	7.73 (10)	7.75 (16)	7.71 (6)
3	7.70 (3.5)	7.74 (13.5)	7.77 (18)	7.74 (13.5)
4	7.70 (3.5)	7.74 (13.5)	7.78 (20)	7.79 (22)
5	7.72 (8)	7.78 (20)	7.80 (23.5)	7.81 (26)
6	7.73 (10)	7.78 (20)	7.81 (26)	7.85 (29)
7	7.73 (10)	7.80 (23.5)	7.84 (28)	7.87 (30)
8	7.76 (17)	7.81 (26)	7.91 (31.5)	7.91 (31.5)
Rank totals (R)	55	132.5	176.5	164
Rank means	6.88	16.56	20.50	22.06

Here,

H_0 is pH of four ponds are not different

H_A is the four ponds differ in pH

As pH is the negative logarithmic value of the hydrogen ion (H^+) concentration, it is not an absolute number; therefore, it shouldn't be averaged as such. These values can be converted to absolute numbers for analysis, but it is a cumbersome task. Instead, a nonparametric test can be performed by using ranks.

Here,

Total no. of data (N) = $8 \times 4 = 32$

Number of treatments (n) = 4 (ponds as treatment)

Rank totals (values in parentheses are ranks):

R1 (Pond 1) = $1 + 2 + 3.5 + 3.5 + 8 + 10 + 10 + 17 = 55$

R2 (Pond 2) = 132.5

R3 (Pond 1) = 176.5

R4 (Pond 1) = 164

$$H = 12/[N(N+1)] \times \Sigma\ R^2/n - 3(N+1)$$
$$= 12/(32 \times 33) \times [55^2 + 132.5^2 + 145^2 + 164^2]/4 - 3 \times (32+1)$$
$$= 12.69$$

Correction factor (C) = $1 - \Sigma T/(N^3 - N)$

T – tied group, check no. of tied groups:

$3.5(3, 4), 6(5, 6, 7), 10(9, 10, 11), 13.5(12, 13, 14, 15)$

$20(19, 20, 21), 23.5(23, 24), 26(25, 26, 27), 31.5(31, 32)$

Sum of tied groups (ΣT)

$$= \Sigma(t^3 - t)$$
$$= (2^3 - 2) + (3^3 - 3) + (3^3 - 3) + (4^3 - 4) + (3^3 - 3) + (2^3 - 2)$$
$$+ (3^3 - 3) + (2^3 - 2)$$
$$= 174$$

Table 7.8 Pair-wise comparisons between two ponds at a time.

Comparisons	Difference in rank means (d)	SE $= d/SEp$	$t_{0.05,7}$	Statistical Inference
Pond 1 vs. Pond 2	9.69	2.07	2.639	NS
Pond 1 vs. Pond 3	15.19	3.25	2.639	S
Pond 1 vs. Pond 4	13.63	2.91	2.639	S
Pond 2 vs. Pond 3	5.50	1.18	2.639	NS
Pond 2 vs. Pond 4	3.94	0.84	2.639	NS
Pond 3 vs. Pond 4	1.56	0.33	2.639	NS

$$\text{Correction factor } (C) = 1 - \Sigma T/(N^3 - N)$$
$$= 1 - 174/(32^3 - N)$$
$$= 0.995$$

$Hc = H/C = 12.69/0.995 = 12.76$ (Check the table of x^2)
Tabulated value $(x^2_{0.05,3}) = 7.815$

Calculated x^2 values (12.76) are higher than the standard value (7.815) for 3 df at 5% level of significance. Therefore, H_0 is rejected, which means pH values are significantly ($P < 0.05$) different among the ponds. However, it does not indicate which ponds differ with which ones. For this purpose, as in parametric test, multiple comparisons among ponds need to be performed by using mean ranks rather than original data. The same data set is used for multiple comparisons. The basic principle is that the difference between two pairs of means is converted to the standard difference as $d = (A - B)/SE$.

For the Kruskal–Wallis test, SE is calculated as:

$$SEp = \sqrt{[N(N + 1)/12 - \Sigma T/\{12(N - 1)\}] \times 2/n}$$
$$= \sqrt{[32 \times (33)/12 - 174/(12 \times 31)] \times 2/8} = 4.68$$

As with the parametric test, the final result can be presented either in tabular form (Table 7.9) or in graphical form as mentioned/shown above.

Based on the results, it can be concluded that the pH of Pond 1 was significantly lower than in Ponds 3 and 4. Ponds 2, 3, and 4 had no significant differences ($P > 0.05$) in pH. Similarly, Ponds 1 and 2 had no significant difference in pH. In this case, it can't be stated that Pond 3 had the highest pH, nor can be said Pond 1 had the lowest pH.

Table 7.9 Comparison of means based on their ranks.

	Pond 1	Pond 2	Pond 3	Pond 4
Mean Ranks	6.88	16.56	20.50	22.06

7.3   Randomized complete block design

The randomized complete block design (RCBD) is probably the most widely used design because, in reality, it is difficult to find all identical or uniform experimental units in the field of aquaculture, especially in outdoor ponds. Some of them are closer to or separated by canals, roads, shade, etc. Even when using cages, some of them are closer to dikes, whereas others can be far away. Similarly, few rows of indoor tanks can be in a darker area, whereas others can be in brighter areas. These factors can have large effects on response variables, but these effects can neither be avoided nor even minimized to negligible levels. In such cases, the only option is to separate their effects while designing the experiment by blocking. The experimental units that are thought to be uniform are considered one block. Blocking minimizes the random error by separating the experimental/random error, thereby maximizing the chance of treatment effects becoming significant. However, care should be taken while designing the experiment. All of the treatments have to be included in each block. Therefore, each block needs to be subdivided into the experimental units equal to the number of treatments. Treatments are completely randomized in each block, which means that a block is a single replication of the experiment. Using the RCBD, the resulting ANOVA can separate variations due to treatments, blocks, and residual error. Appropriate blocking is to minimize the variance among experimental units within blocks while maximizing the variation among blocks. Precision usually decreases as the number of experimental units/treatments and the size of units per block increase. Therefore, block size or the number of treatments should be kept as small as possible. The following steps are followed for randomization:

1. Determine the total number of experimental units, $n =$ treatments $(t) \times$ blocks (b), as an example shown in Table 7.10.
2. Assign all of the treatments randomly (e.g. Figure 7.4) in each block by using a lottery or random table as described for CRD.

Model for a RCBD:

$$Y_{ij} = m + T_i + B_j + R_{ij}$$

Table 7.10 Experimental design for 6 treatments $\times$ 3 blocks.

Block	T1	T2	T3	T4	T5	T6
1	T1B1	T2B1	T3B1	T4B1	T5B1	T6B1
2	T1B2	T2B2	T3B2	T4B2	T5B2	T6B2
3	T1B3	T2B3	T3B3	T4B3	T5B3	T6B3

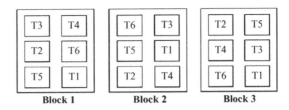

Figure 7.4 All six treatments are randomized within each block.

Where,

Y_{ij} is the observed value for the jth replicate of the ith treatment (where $i = 1$ to t and $j = 1$ to n)

m   is the grand mean

T_i is the treatment effect for the ith treatment; the treatment effects may be either fixed or random.

B_j is the block effect for the jth block; the block effect may be either fixed or random; however, if treatments are fixed, then random blocks are required for exact tests of treatment hypotheses.

R_{ij} is the random error associated with the Y_{ij} experimental unit

If $T_i > R_{ij}$, then treatment effect is significant after separation of block effects.

It is clear from Figure 7.5 that two null hypotheses (H_0) are tested in this design, i.e. there is no effect of block and no effect of treatments; therefore, it is called a two-way ANOVA.

7.3.1   Parametric test using ANOVA

1. Group the data by treatments and calculate the treatment totals (T), block totals (B), and grand total (G), grand mean, and coefficient of variation (CV), etc.
2. Using the number of treatments (t) and the number of blocks (b), determine the df for each source of variation.

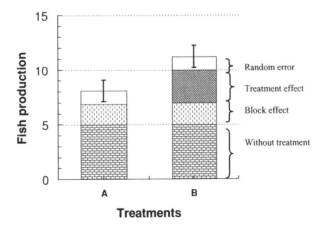

Figure 7.5 Separation of effects of treatment, block, and random error.

Table 7.11 ANOVA table for RCBD.

Source of variation	Sum of squares	df.	Mean square	F	p
Trts	SS_T	$t-1$	$\dfrac{SS_T}{t-1} = MST$	$\dfrac{MST}{MSE}$	$p < 0.05$ * $P < 0.01$ ** $P > 0.05$ ns
Block	SS_B	$b-1$	$\dfrac{SS_B}{b-1} = MSB$	$\dfrac{MSB}{MSE}$	$p < 0.05$ * $P < 0.01$ ** $P > 0.05$ ns
Residual error	SS_E	Total – others	$\dfrac{SS_E}{df\,(E)} = MSE$		
Total	TSS	$(t \times b) - 1$			

3. Construct an outline/table of the ANOVA, as shown in Table 7.11.
4. Using X_i to represent the measurement of the ith plot, T_i as the total of the *i*th treatment, and n as the total number of experimental plots [i.e. $n = rt$], calculate the correction factor (CF) and the various sums of square (SS).
5. Calculate the mean square (MS) for each source of variation by dividing SS by their corresponding df.
6. Calculate the F-values (R.A. Fisher) for testing significance of the treatment and block differences, i.e. mean square of treatment divided by the mean square of the error ($F = MST/MSE$) and mean square of block divided by the mean square of the error (MSB/MSE).
7. Enter all of the computed values in the ANOVA table.
8. If there is an effect of block, then treatment effects need to be compared within each block separately.
9. If there is no significant block effect, data can be analyzed using one-way ANOVA and then t-test to compare two means and multiple range tests to compare among means at a given time. DMRT is the most common, although Tukey's HSD has been suggested as middle path as other tests are considered either too conservative or too moderate.
10. If ANOVA shows no significant difference, then multiple range tests are not necessary.

For the purpose of ease of describing the method, the same example for CRD (Table 7.4) is used here, assuming that the feeds were tested in five different ponds instead of a single large pond as in CRD. For example, 4 iso-protein commercial catfish pellets differing in lipid levels were compared, stocking 50 fish in each of the cages installed in 5 ponds. Data shown below are the mean final weights (g) of fish (fish from a cage in Pond 5 assigned for Feed 3 escaped during handling). Test whether these feeds (treatment) and ponds show any significant differences in terms of fish growth.

Here, $H_0 = m_1 = m_2 = m_3 = m_4$ and there is no effect of pond (block).

Step 1: The missing value can be either estimated or left empty. Researchers have been known to use zero (0) in place of the missing value, which is wrong.

Table 7.12 Mean final weights of fish with an estimated value for Feed 3 in Pond 5.

Pond (Block)	Feed 1	Feed 2	Feed 3	Feed 4	Total
1	86	88	105	94	374
2	83	87	104	91	365
3	91	94	102	89	376
4	84	86	99	92	361
5	87	89	105	97	378
Total	431	444	515	463	1,853
n	5	5	5	5	20
Mean	86.2	88.8	103.0	92.5	

Zero value means there is no production at all, or all fish have died in the case of survival data. A zero value drastically lowers the mean and increases the variance, leaving little space for detecting the treatment effects, i.e. increased chance of committing Type II error. It is better to leave the missing value as blank and use one less df. This could be done here as well, but for the sake of describing the method, the missing value has been estimated as:

Missing value = $[(t \times$ treatment total $+ b \times$ block total$) -$ grand total$]/(t-1)$
  $(b-1)$
  $= [(4 \times 410 + 5 \times 273) - 1748]/(3 \times 4) = 105$

After filling in the estimated missing value in Table 7.4 for the purpose of ANOVA, Table 7.12 has been presented.

If there are more than two missing values, one of them should be guessed and the other is estimated by using the guessed value. Then, the guessed value is estimated back by using the estimated value. Closer values can be obtained by iterating until both of them become stable. Although it is difficult, more than two missing values are possible to estimate. However, as a general rule, missing values should not be more than 10%, and they should be used only when data are lost due to unavoidable circumstances, not as an escape for proper management of the research trial or survey.

Step 2: Calculate sum of squares.
Correction factor (C) = (grand total)$^2/n$ = $(1853)^2/20$
  $= 171,680$
Total SS $= (86)^2 + (88)^2 + \cdots + (97)^2 - C = 954$
Treatment SS $= \Sigma$(treatment total)$^2/t - C$
  $= (431)^2/5 + 444^2/5 + 515^2/5 + 463^2/5 - 172,634$
  $= 821$
Block SS $= \Sigma$(block total)$^2/n - C$
  $= (374)^2/4 + (365)^2/4 + (376)^2/4 + (361)^2/4 + (378)^2/4 - 172,634$
  $= 57$
Error SS $=$ total SS $-$ SST $-$ SSB
  $= 954 - 821 - 57 = 76$

Table 7.13 ANOVA table for RCBD.

Sources of Variation	SS	df	Mean SS = SS/df	F = MST/MSE	P	Significance
Total	954	19			<0.01	**
Treatment	821	3	273.6	43.2	>0.05	ns
Block	57	4	14.3	2.3		
Error	76	12	6.3			

Note: Numerator $df = 3$; denominator $df. = 12$ for treatment F and 4 and 12, respectively, for block F. In most statistical packages, correction factor is called as "intercept."

Step 3: Prepare an ANOVA table as shown in Table 7.13.

Here, statistical inference is "reject H_0" for treatment but "accept H_0" for block, which means the type of feed has a highly significant ($P < 0.01$) effect, but pond has no effect ($P > 0.05$) on fish growth. Therefore, the analysis can actually be done as in the case of CRD, and further comparison among means is done by using multiple comparisons as shown in CRD (Section 7.2.1). However, if the pond (block) had significant effects, multiple comparisons couldn't be performed on pooled means of all five ponds. Instead, the five ponds used for the trial would be compared by using the multiple comparison method in order to locate the pairs of ponds having differences. It could even be possible to determine which pond had the best fish growth and which one had the lowest. Therefore, if there was a need to select the best ponds for growth, it could be possible.

Other examples of RCBD

1. Efficacy of different drugs on different age or strain groups of fish in which age or strain is considered as block.
2. Effects of feeding rate on tilapia seed output harvested weekly, where time should be considered as block and feeding rate as treatment.
3. Fish growth trial in cages or hapas in ponds, where ponds can be blocks.
4. Comparison among organic, inorganic, and their combination on the fish productivity in different agroecological contexts.

7.3.2   Nonparametric test: Friedman test for RCBD

As an example, a data set of daily weight gain of fish has been used to describe the method of Friedman test as shown in Tables 7.14 and 7.16. With this method, treatment effect and block effect are tested separately because ranking must be done separately.

7.3.2.1   Testing treatment effect

$$x^2 = [12\Sigma\ R_i^2/ab(b+1)] - 3a(b+1)$$

Table 7.14 Daily weight gain (g) of fish, obtained from a trial.

Block (b)	Diets (a)			
	1	2	3	4
1	1.5 (2)	2.7 (4)	2.1 (3)	1.3 (1)
2	1.4 (2)	2.9 (4)	2.2 (3)	1.0 (1)
3	1.4 (2)	2.1 (3)	2.4 (4)	1.1 (1)
4	1.2 (1)	3.0 (4)	2.0 (3)	1.3 (2)
5	1.4 (1)	3.3 (4)	2.5 (3)	1.5 (2)
Rank sum (R_i)	8	19	16	7

Where,

    a   is treatment,

    b   is block,

$$x^2 = 12/[5 \times 4(4 + 1)][12.5^2 + 10^2 + 10^2 + 18^2 + 9.5^2] - 3$$
$$\times 5(4 + 1)] = 12.6$$
$$v = a - 1 = 4 - 1 = 3$$

$x^2_{0.05,3} = 7.815$ (From x^2 table, Appendix A2)

Reject H_0 ($P < 0.05$), which means weight gain of fish was significantly different or diet has a significant effect; then, proceed to multiple comparisons test.

$$SE = ba(a + 1)/12 = 5 \times 4 \times (4 + 1)/12 = 2.89$$

From table, $Q_{0.05, 3} = 2.639$

Therefore, critical difference (*d*) = $2.89 \times 2.394 = 6.9$

Multiple comparison results, presented in Table 7.15, showed that Diets 2 and 3 gave significantly higher daily weight gain of fish as compared with Diets 1 and 4. But there was no significant difference between Diets 1 and 4 and between Diets 2 and 3.

7.3.2.2   Testing block effect

For the test of block effect, ranking is done within the treatment for each diet, as shown in Table 7.16.

$$x^2 = [12 \, \Sigma \, R_i^2/ab(b + 1)] - 3a(b + 1)$$

Table 7.15 Multiple comparison among treatments.

	Diets (a)			
Block (*b*)	4	1	3	2
Rank sum (R_i)	7	8	16	19

Table 7.16 Daily weight gain (g) of fish, obtained from a trial.

Block (*b*)	Diets (*a*)				Block Sum (R_j)
	1	**2**	**3**	**4**	
1	1.5 (5)	2.7 (2)	2.1 (2)	1.3 (3.5)	12.5
2	1.4 (3)	2.9 (3)	2.2 (3)	1.0 (1)	10
3	1.4 (3)	2.1 (1)	2.4 (4)	1.1 (2)	10
4	1.2 (1)	3.0 (4)	2.0 (1)	1.3 (3.5)	9.5
5	1.4 (3)	3.3 (5)	2.5 (5)	1.5 (5)	18

Where,

- *a* is treatment,
- *b* is block

$$x^2 = 12/[5 \times 4(4+1)][12.5^2 + 10^2 + 10^2 + 9.5^2 + 18^2] - 3 \times 5(4+1)]$$
$$= 17.46$$
$$v = a - 1 = 5 - 1 = 4$$
$$x^2_{0.05,\ 4} = 9.49 \text{ (from table)}$$

Reject N_0 ($P < 0.05$), which means the block has significant effect. Comparisons among the blocks are done as follows:

$$SE = ab(b+1)/12 = 4 \times 5 \times (5+1)/12 = 3.1$$

From table, $Q_{0.05,\ 4} = 2.639$
Therefore, critical difference (*d*) = 3.16 × 2.639 = 8.3
Multiple comparison results are presented in Table 7.17.

  Multiple comparison results showed that Block 5 had a significantly higher daily weight gain of fish as compared with Block 4, but there was no difference with Blocks 1, 2, and 3. Similarly, there were no significant differences among Blocks 1, 2, 3, and 4.

Table 7.17 Multiple comparison among blocks.

	Blocks (*b*)				
Block (*b*)	4	3	2	1	5
Rank sum (R_j)	9.5	10	10	12.5	18

7.4   Latin square design

Latin square design is used where there are two distinct nuisance factors affecting from different directions. However, testing is done for a single factor of primary interest. The design is often called two-way block design and is less common in biological research. In this design, the number of blocks (both sides), number of units within each block, and number of treatments need to be equal. The main benefit of this design is that it reduces the size of the experiment by including one more factor in the two-factor design. For example, if a design has two factors (f1 and f2) with 5 levels each, then the total number of experimental units required will be 25. Without increasing the experimental units, another factor (f3) can be included by assigning its levels randomly in each column and row. Otherwise, addition of the third factor would increase the number of experimental units by 5-fold, i.e. $25 \times 5 = 125$, which is not easy to manage. However, this design is still restricted to few treatments only, normally four to eight. All of the observations are indispensable for analysis in this design; losing a single observation can cause an error. If this occurs due to unavoidable circumstances, the missing value (lost observation) has to be estimated before performing ANOVA, as described in Section 7.3. Three hypotheses are tested by comparing the variation; therefore, it is called multifactor ANOVA. Three null hypotheses would mean there are no effects of Block 1, Block 2, and the treatment.

As it is a unique design, randomization and layout are quite difficult. All of the treatments have to be randomly allocated to each block of both factors. In other words, all the treatments in every block, from both sides, need to be allocated randomly; an example is shown in Figure 7.6.

A model for Latin square block design is below, and the partition of errors or variations is shown in Figure 7.7.

$$Y_{ij} = m + T_i + B_1 + B_2 + R_i$$

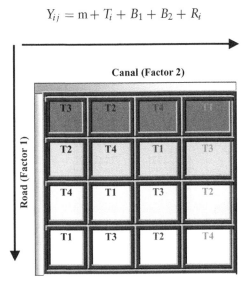

Figure 7.6 Design and layout of treatments in a Latin square design.

Table 7.18 Layout and data arrangement in Latin square design.

Row	Column			
	1	2	3	4
1	T3	T2	T4	T1
2	T2	T4	T1	T3
3	T4	T1	T3	T2
4	T1	T3	T2	T4
Total				
Treatment 1 total				
Treatment 2 "				
Treatment 3 "				
Treatment 4 "				
Grand total "				

Where,

Y_{ij}　is the observed value
m　　is the grand mean
T_i　is the treatment effect
B_1　is the effect of first block
B_2　is the effect of second block
R_i　is the random error

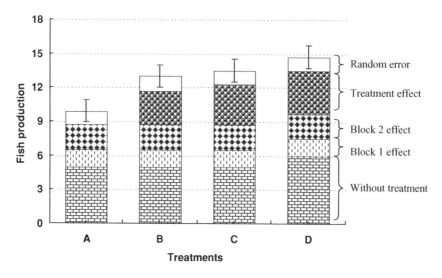

Figure 7.7 Division of variability in a Latin square design.

The following steps are used in estimating and testing a model using ANOVA:

1. Group the data by blocks and treatments and calculate the treatment totals (T), block totals (B), grand total (G), grand mean, and the coefficient of variation (CV), etc. as shown in Table 7.18.
2. Using the number of treatments (t) and the number of blocks (b), determine the df for each source of variation.
3. Construct an outline/table of the ANOVA as shown in Table 7.19.
4. Using X_i to represent the measurement of the ith plot, T_i as the total of the ith treatment, and n as the total number of experimental plots (i.e. $n = rt$), calculate the correction factor (CF) and the various sums of square (SS).
5. Calculate the mean square (MS) for each source of variation by dividing SS by their corresponding df.
6. Calculate the F-values (R.A. Fisher) for testing significance of the treatment difference ($F = $ MST/MSE and MSB/MSE).
7. Enter all of the values computed in the ANOVA table.
8. If there is effect of a block, treatment effects need to be compared within each block of that particular factor.
9. If there are no significant effects of any block, data can be analyzed by using one-way ANOVA and go for t-test and multiple range tests to compare treatments (DMRT and Tukey's HSD are the most common choices, depending on the situation).
10. Note: as in other designs, if ANOVA shows no significant difference, then multiple range tests are not necessary.

Latin square design helps when there are two nuisance factors that either can't be combined into a single factor or we wish to keep separate. However, it has limited use as it allows for a relatively small number of treatments. Sometimes it is quite difficult to have the number of levels of each blocking variable equal to the number of levels of the treatment factor. The Latin square model normally assumes that there are no interactions between the blocking variables or between the treatment variable and the blocking variable.

Table 7.19 ANOVA table for Latin square design.

Source of Variation	SS	df	MS	F	P
Treatments	SS_T	$t - 1$	$\dfrac{SST}{t-1} = MST$	$\dfrac{MST}{MSE}$	$P < 0.05^*$ $P < 0.01^{**}$ $P > 0.05^{ns}$
Row block	SS_R	$t - 1$	$\dfrac{SSR}{t-1} = MSR$	$\dfrac{MSR}{MSE}$	$''$
Column block	SS_C	$t - 1$	$\dfrac{SSC}{t-1} = MSC$	$\dfrac{MSC}{MSE}$	$''$
Residual error	SS_E	total − others	$\dfrac{SSE}{df(E)} = MSE$		
Total	total SS	$t^2 - 1$			

Variances can be computed as described in RCBD. The only difference is that there is more block in this design, which means "error" variance is further split. All of the methods are the same, including the computation; F-values for treatment, Block 1 and Block 2, and their SS are divided by "error MS" separately, as shown in the Table 7.19.

As the manual calculations of variances are quite complicated and time consuming, and because there are several statistical software packages readily and easily available, we suggest that researchers use them. The following example is given to describe the method for the guidance purpose:

An experiment was conducted to compare growth performance of four strains (A, B, C, and D) of Nile tilapia being fed four types of feeds (Feed 1, 2, 3, and 4) in four ponds (Ponds 1, 2, 3, and 4). All of the fish were from a uniform population randomly drawn and allocated to each tank for the trial. Fish were grown for 3 months and fed at 2% biomass per day. Table 7.20 presents a summary of the data collected.

Here,

H_{01} → there is no difference among the tilapia strains.

H_{02} → there is effect of feed types.

H_{03} → there is no effect of pond.

As the treatment factor is randomized in both directions, treatment means must be computed separately, as shown in Table 7.20. Statistical analysis can be performed by using various statistical software programs, which are readily available. Data arrangement is similar in most of the programs; shown in Table 7.21. If data have been arranged taking only one variable, the method is called univariate analysis. More columns can be added on the right if more variables have been collected for analysis. For description purposes, we have used only one variable.

Table 7.22 shows the results of univariate ANOVA. Out of the three sources of variations (strain, feed, and pond), only the first two have statistically significant effects on the final weight of the fish, which is confirmed from P (Sig.) values. As it was tested using 0.05 level of significance, the factors which have less than

Table 7.20 Final weights (g) of Nile tilapia at harvest.

Pond	Feed 1	Feed 2	Feed 3	Feed 4	Total	Mean	SE
1	170.6 (C)	198.5 (D)	204.8 (C)	214.3 (D)	788.2	197.1	9.4
2	160.6 (A)	187.2 (C)	212.2 (D)	200.5 (B)	760.5	190.1	11.1
3	190.6 (D)	184.5 (B)	202.4 (B)	196.4 (A)	773.9	193.5	3.8
4	165.6 (B)	159.5 (A)	194.7 (A)	202.0 (C)	721.8	180.5	10.5
Total	687.4	729.7	814.1	813.2	3,044.4	190.3	3.6
Mean	171.9	182.4	203.5	203.3			
	Strain A	Strain B	Strain C	Strain D			
Total	711.2	753.0	764.6	815.6			
Mean	177.8	188.3	191.2	203.9			
SE	10.26	7.47	8.97	5.65			

Table 7.21 Data arrangement for statistical analysis using computer software.

Pond (Random Factor1)	Feeding Regime (Fixed Factor1)	Strain (Fixed Factor2)	Final Weight (Independent Variable)
1	1	3	170.6
2	1	1	160.6
3	1	4	190.6
4	1	2	165.6
1	2	4	198.5
2	2	3	187.2
3	2	2	184.5
4	2	1	159.5
1	3	3	204.8
2	3	4	212.2
3	3	2	202.4
4	3	1	194.7
1	4	4	214.3
2	4	2	200.5
3	4	1	196.4
4	4	3	202.0

0.05 (5%) *P* value (Sig.) are significantly affecting the dependent variable (final weight of fish). This means that the probability of these factors affecting the weight of fish by chance is less than 5%, or it is 95% sure that the effect is due to these factors. Although pond (block) was also thought to be a random factor while designing the trial, it did not actually make any significant impact as the *P* value (Sig. = 0.296) is higher than 0.05. This means that only $100 - 29.6 = 70.4\%$ of the variation in fish weight is due to the pond, not 95%. Therefore, the effect of the pond is insignificant.

After pointing out which factors affect the dependent variable (fish weight), it is necessary to compare among the levels of each factor by using multiple comparisons, especially when researchers need to find the best ones. For example,

Table 7.22 Tests of Between-Subjects Effects: dependent variable: weight.

Sources		Type III SS	df	MS	F	Sig. (P)	Partial Eta Squared	Observed Power[a]
Intercept	Hypothesis	579273	1	579273	13290	0.000	1.000	1.000
	error	104	2	43.587[b]				
Strain	Hypothesis	896	3	298.6	11.4	0.007	0.851	0.947
	error	157	6	26.229[c]				
Feed	Hypothesis	2985	3	995.1	37.94	0.000	0.950	1.000
	error	157	6	26.229[c]				
Pond	Hypothesis	122	3	40.69	1.55	0.296	0.437	0.238
	error	157	6	26.229[c]				

[a] Computed using alpha = 0.05.
[b] 1.200 MS(Pond) – 0.200 MS(Error).
[c] MS(Error).

the best strain of fish or feed has to be selected to make a recommendation. For this purpose, multiple comparisons, often called post-hoc test, are performed, which are available with most statistical software. Examples are shown in Tables 7.23 and 7.24.

Table 7.23 Multiple comparisons using Tukey's HSD post-hoc tests for dependent variable: weight.

(*I*) Feed	(*J*) Feed	Mean Difference (*I* − *J*)	SE	Sig.	95% CI Lower Bound	95% CI Upper Bound
1.00	2.00	−10.5750	3.62139	0.095	−23.1112	1.9612
	3.00	−31.6750*	3.62139	0.001	−44.2112	−19.1388
	4.00	−31.4500*	3.62139	0.001	−43.9862	−18.9138
2.00	1.00	10.5750	3.62139	0.095	−1.9612	23.1112
	3.00	−21.1000*	3.62139	0.005	−33.6362	−8.5638
	4.00	−20.8750*	3.62139	0.005	−33.4112	−8.3388
3.00	1.00	31.6750*	3.62139	0.001	19.1388	44.2112
	2.00	21.1000*	3.62139	0.005	8.5638	33.6362
	4.00	0.2250	3.62139	1.000	−12.3112	12.7612
4.00	1.00	31.4500*	3.62139	0.001	18.9138	43.9862
	2.00	20.8750*	3.62139	0.005	8.3388	33.4112
	3.00	−0.2250	3.62139	1.000	−12.7612	12.3112

Based on observed means.
* The mean difference is significant at the 0.05 level.

Multiple comparisons show the combination of each pair and their probabilities (P = Sign.). At the same time, most statistical software shows homogeneous subsets (Table 7.25), which makes it easier to understand. For example, the mean weights of fish weight appearing in the same subset are not significantly

Table 7.24 Multiple comparisons using Tukey's HSD post-hoc tests for dependent variable: strain.

(*I*) Strain	(*J*) Strain	Mean Difference (*I* − *J*)	SE	Sig.	95% CI Lower Bound	95% CI Upper Bound
1.00	2.00	−10.4500	3.62139	0.099	−22.9862	2.0862
	3.00	−13.3500*	3.62139	0.039	−25.8862	−0.8138
	4.00	−26.1000*	3.62139	0.001	−38.6362	−13.5638
2.00	1.00	10.4500	3.62139	0.099	−2.0862	22.9862
	3.00	−2.9000	3.62139	0.852	−15.4362	9.6362
	4.00	−15.6500*	3.62139	0.019	−28.1862	−3.1138
3.00	1.00	13.3500*	3.62139	0.039	0.8138	25.8862
	2.00	2.9000	3.62139	0.852	−9.6362	15.4362
	4.00	−12.7500*	3.62139	0.047	−25.2862	−0.2138
4.00	1.00	26.1000*	3.62139	0.001	13.5638	38.6362
	2.00	15.6500*	3.62139	0.019	3.1138	28.1862
	3.00	12.7500*	3.62139	0.047	0.2138	25.2862

Based on observed means.
* The mean difference is significant at the 0.05 level.

Table 7.25 Homogeneous subsets of weight using feed as factor (Tukey's HSD test).

| Feed | N | Subsets | |
		1	2
1	4	171.8500	
2	4	182.4250	
4	4		203.3000
3	4		203.5250
Sig.		0.095	1.000

Notes: means for groups in homogeneous subsets are displayed;. based on Type III SS; the error term is MS(Error) = 26.229 and the level of Significance (Sig.) i.e. Alpha = 0.05.

different, e.g. Feeds 1 and 2 do not differ. Similarly, Feeds 3 and 4 appearing in Subset 2 also did not differ. Feeds 3 and 4 produce significantly weightier fish than Feeds 1 and 2. In the case of strain (Table 7.26), results show that the weight of Strain 4, which appear in the third subset alone, is considered the highest. Strain 3 appeared with Strain 2 in Subset 2, which means they do not differ; but Strain 3 differs from Strain 1. Strains 1 and 2 do not differ as they appear together in Subset 1. In this case, Strain 2 appears both in Subset 1 as well as in Subset 2, which means the weight of Strain 2 does not differ from all of the weights appearing in both subsets.

These results can also be presented in graphical form, presented in a more attractive and clear way, using error bars (SD or SE) and alphabetical notations. These error bars show the variations of the means, which means they can go up and down up to those levels. They roughly tell us that, if the error bars created from standard errors of two means are overlapping, the two means do not significantly differ. At the same time, if the two means do not have overlapping error bars, there is a chance of them being significantly different. Therefore, it is a good method of presenting means with some statistical sense. More importantly, alphabets are used to show the confirmation of the statistical differences. The bars or the means with the same alphabetical notations are not significantly

Table 7.26 Homogeneous subsets of weight using fish strain as factor (Tukey's HSD test).

| Strain | N | Subsets | | |
		1	2	3
1	4	177.8000		
2	4	188.2500	188.2500	
3	4		191.1500	
4	4			203.9000
Sig.		0.099	0.852	1.000

Notes: means for groups in homogeneous subsets are displayed; based on Type III SS; the error term is MS(Error) = 26.229 and the level of Significance (Sig.) i.e. Alpha = 0.05.

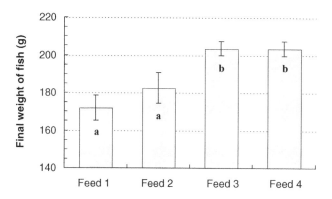

Figure 7.8 Comparison among feeds for the weight of fish.

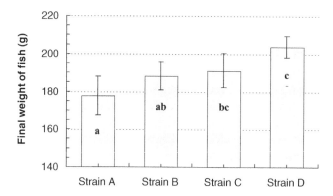

Figure 7.9 Comparison among strains based on the growth of fish.

different. Conversely, the means or the bars not having the same alphabets are significantly different, which is shown in Figures 7.8 and 7.9.

7.5   Factorial experiments

In actual biological systems, organisms are exposed to many factors simultaneously. Response to one factor may vary with the response to the levels of other factors. A single-factor experiment is the simplest method of carrying out research, but this alone can't explain complicated biological systems where there are many factors interacting. Therefore, factorial designs are particularly useful in which two or more than two fixed factors (treatment factors), which have graded levels, are tested at one time. Depending on the number of factors and their levels, there are several types of factorial experiments, which are shown in Table 7.27.

Table 7.27 Types of factorial design.

Factors	Level	Factorial Designs	No. of Treatments	Replicates	No. of Experimental Units Required
Two factors	Same	2 × 2	4	3	12
	Different	2 × 3	6	3	18
Three factors	Same	2 × 2 × 2	8	3	24
		2 × 2 × 3	12	4	48
	Different	4 × 3 × 2	24	3	72
		4 × 4 × 2	32	3	96

2 × 2 factorial designs

The simplest factorial design is 2 × 2 (two factors with two levels), in which this design effect of main factors (e.g. effects of Factor *A* and Factor *B* separately) as well as effect due to the interactions between factors (e.g. *A* × *B* or *N* × *P*) are tested. The interactions between two factors are shown in Figures 7.10 and 7.11.

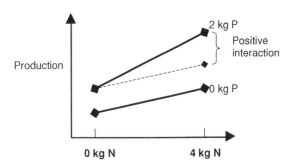

Figure 7.10 Positive interaction between nitrogen and phosphorous fertilization.

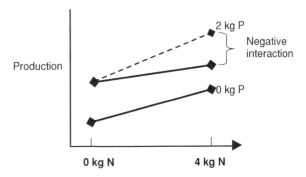

Figure 7.11 Negative interaction between nitrogen and phosphorous fertilization.

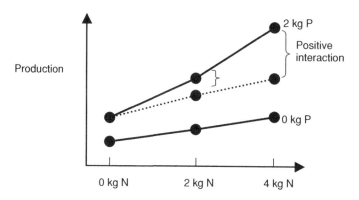

Figure 7.12 Positive interactions between nitrogen and phosphorous fertilization for three application levels of nitrogen.

3×2 factorial designs are shown in Figure 7.12.

The ANOVA model for a two-factor model is shown in Figures 7.13 and 7.14 and Table 7.28:

$$Y_i = m + A + B + (A \times B) + R_i$$

There are three types of models, depending on the type of factors involved, which is shown in Table 7.29. In Model I, only treatment or fixed factors are involved, whereas in Model II, only random or blocks are involved. Model III includes both factors. The method of analysis differs slightly as the mean squares are divided by either the error MS or interaction MS. Table 7.29 summarizes the computation of F-values.

Three-factor factorial design:

Layout and randomization for a three-factor factorial designis done similarly to the CRD and RCBD shown in 2×2 factorial designs in Figure 7.13 by adding more factors and interactions among themselves, for example:
H₀: Null hypotheses

1. There is no effect of factor *A*.
2. There is no effect of factor *B*.
3. There is no effect of factor *C*.
4. There is no interaction effect of *A* & *B*.
5. There is no interaction effect of *A* & *C*.
6. There is no interaction effect of *B* & *C*.
7. There is no interaction effect of *A*, *B*, & *C*.
8. There is no effect of block (if designed with block).

ANOVA models (Table 7.30)
Without Block:

$$Y_i = m + A + B + C + AB + AC + BC + ABC + R_i$$

2 x 2 Factorial in CRD

2 x 2 Factorial in RCBD

Layout and randomization

T2	T3	T4	T1
T4	T1	T3	T2
T3	T1	T2	T4

Block 1

T2	T3	T4	T1

Block 2

T4	T1	T3	T2

Block 3

T3	T1	T2	T4

Null hypothesis

H_0: There is no effect of factor A
There is no effect of factor B
There is no interaction effect of A & B

H_0: There is no effect of factor A
There is no effect of factor B
There is no interaction effect of A & B
There is no effect of block

Figure 7.13 Layout and randomization of treatments in CRD (left) and RCBD (right).

With Block:

$$Y_i = m + A + B + C + AB + AC + BC + ABC + Block + R_i$$

Factorial design is quite common in aquaculture, and an understanding of interaction is important. A following simple example follows.

Effects of vitamin C (mg·kg^{-1} diet) and crude protein (%) levels were tested to determine the effects on the weight of fish using factorial design (Table 7.31). Five levels of vitamin C and three levels of crude protein are considered, which forms a 5×3 factorial design, resulting in 15 treatments. These treatments are

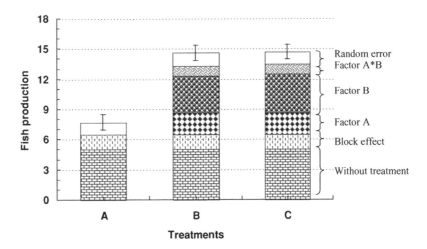

Figure 7.14 Separation variability in factorial design.

Table 7.28 Two-factor ANOVA table.

Source of Variation	SS	df	MS	F	P
Main effects	SST	$ab - 1$	$\dfrac{SST}{ab - 1} = MST$	Depends on model	$P < 0.05$ * $P < 0.01$ ** $P > 0.05$ ns
Factor A	SSA	$a - 1$	$\dfrac{SSA}{a - 1} = MSA$	Depends on model	"
Factor B	SSB	$b - 1$	$\dfrac{SSB}{b - 1} = MSB$	Depends on model	"
Interaction AB	SSAB	$(a - 1)(b - 1)$	$\dfrac{SSB}{(a - 1)(b - 1)} = MSAB$	Depends on model	"
Residual error	SSE	Total − others	$\dfrac{SSE}{df(E)} = MSE$		
Total	Total SS	$abr - 1$			

Table 7.29 ANOVA models.

Hypothesized Effects	Model I (A & B both are treatments)	Model II (A & B both are blocks)	Model III (A – Treatment & B – block)
Factor A	$\dfrac{\text{Factor A MS}}{\text{Error MS}}$	$\dfrac{\text{Factor A MS}}{\text{AB MS}}$	$\dfrac{\text{Factor A MS}}{\text{AB MS}}$
Factor B	$\dfrac{\text{Factor B MS}}{\text{Error MS}}$	$\dfrac{\text{Factor B MS}}{\text{AB MS}}$	$\dfrac{\text{Factor B MS}}{\text{Error MS}}$
AB interaction	$\dfrac{\text{AB MS}}{\text{Error MS}}$	$\dfrac{\text{AB MS}}{\text{Error MS}}$	$\dfrac{\text{AB MS}}{\text{Error MS}}$

Notes: if there is no interaction, AB MS and the Error MS will be the same.

Table 7.30 ANOVA table for factorial design with three factors.

Source of Variation	SS	df	MS	F	P
Main effects	SST	$abc - 1$	$SST/(ab - 1) = MST$		
Factor A	SSA	$a - 1$	$SSA/(a - 1) = MSA$		
Factor B	SSB	$b - 1$	$SSB/(b - 1) = MSB$		
Factor C	SSC	$c - 1$	$SSC/(c - 1) = MSC$		
Interaction AB	SSAB	$(a - 1)(b - 1)$	$SSAB/df = MSAB$		
Interaction AC	SSAC	$(a - 1)(c - 1)$	$SSAC/df = MSAC$		
Interaction BC	SSBC	$(b - 1)(c - 1)$	$SSBC/df = MSBC$		
Interaction ABC	SSABC	$(a - 1)(b - 1)(c - 1)$	$SSABC/df = MSABC$		
Residual error	SSE	Total − others	$SSE/df(E) = MSE$		
Total	Total SS	$abcn - 1$			

Note: The F-values are calculated as corresponding MS values divided by either Error MS (MSE) of Interaction MS (e.g. AB MS, AC MS or BC MS) depending upon the model type as shown in Table 7.29. Their corresponding P-values can be determined using standard tables or software to confirm whether their effects are significance or not.

Table 7.31 A trial with vitamin C and crude protein on the growth of fish.

Crude Protein (%)	Vitamin C (mg·kg^{-1} diet)				
	Level 1: 100 mg	Level 2: 80 mg	Level 3: 60 mg	Level 4: 50 mg	Level 5: 40 mg
20	P20C1	P20C2	P20C3	P20C4	P20C5
30	P30C1	P30C2	P30C3	P30C4	P30C5
40	P40C1	P40C2	P40C3	P40C4	P40 C5

randomly allotted in each of the three different ponds. The resulting factorial experiment in RCBD is shown in Table 7.31.

Where,

Crude protein levels: 20, 30, and 40

Vitamin levels: 1 i.e. control (100 mg·kg^{-1}), 2 (80 mg·kg^{-1}), 3 (60 mg·kg^{-1}), 4 (50 mg·kg^{-1}), and 5 (40 mg·kg^{-1}) All of these treatment combinations are randomized in each pond, which means that each pond will receive all of the treatment combinations, as shown in Figure 7.15.

For the purpose of analysis, Table 7.32 needs to be reconstructed into three two-way tables (Table 7.33a, b, and c).

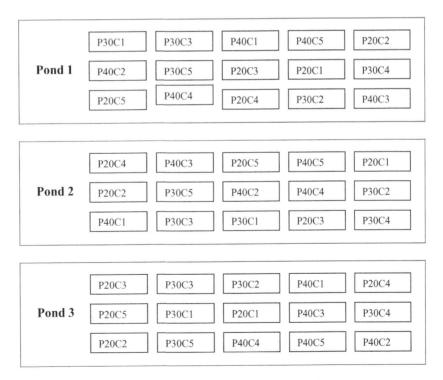

Figure 7.15 Experimental lay-out for a factorial design in three blocks (ponds).

Table 7.32 Final weight of fish (g) at the end of trial.

				Vitamin C Levels			
		C1	C2	C3	C4	C5	Total
Pond 1	Protein 20	648	483	422	419	313	2,284
	Protein 30	793	528	445	438	329	2,532
	Protein 40	873	548	480	409	319	2,629
	Total	2,313	1,559	1,347	1,266	960	7,446
Pond 2	Protein 20	396	332	274	284	191	1,477
	Protein 30	504	354	316	290	236	1,700
	Protein 40	660	373	258	278	230	1,799
	Total	1,560	1,059	848	852	657	4,976
Pond 3	Protein 20	247	197	156	159	100	859
	Protein 30	403	169	159	162	144	1,037
	Protein 40	444	290	194	159	162	1,249
	Total	1,093	656	509	481	406	3,145
							15,567

Calculations are shown below:

1. Grand total $= 15,567$
2. Sum of squares of all observations $= 648^2 + 483^2 + \cdots + 162^2 = 6,714,259$

Table 7.33 Two-way ANOVA tables for factorial design.

a) Two-way table for Pond × vitamin C.

	C1	C2	C3	C4	C5	Sum
Pond 1	2,313	1,559	1,347	1,266	960	7,446
Pond 2	1,560	1,059	848	852	657	4,976
Pond 3	1,093	656	509	481	406	3,145
Sum	4,966	3,274	2,704	2,599	2,023	15,567

b) Two-way table for Pond × Protein.

	Protein20	Protein30	Protein40	Sum
Pond 1	2,284	2,532	2,629	7,446
Pond 2	1,477	1,700	1,799	4,976
Pond 3	859	1,037	1,249	3,145
Sum	4,620	5,270	5,678	15,567

c) Two-way table for Protein × vitamin C.

	C1	C2	C3	C4	C5	Sum
Protein 20	1,290	1,012	853	862	604	4,620
Protein 30	1,700	1,051	920	891	708	5,270
Protein 40	1,977	1,211	932	846	711	5,678
Total	4,966	3,274	2,704	2,599	2,023	15,567

3. Sum of squared totals for pond/sample size $= (7,446^2 + 4,976^2 + 3,145^2)/(3 \times 5) = 6,006,526$

4. Sum of squared totals for vitamin C/sample size $= (4,966^2 + 3,274^2 + 2,704^2 + 2,599^2 + 2,023^2)/(3 \times 3) = 5,949,866$

5. Sum of squared totals for protein/sample size $= (4,620^2 + 5,270^2 + 5,678^2)/(5 \times 3) = 5,423,355$

6. Sum of squared cell totals/sample size (Pond $\times$ vitamin C, two-way table) $= (2,313^2 + 1,559^2 + \ldots + 481^2 + 406^2)/3 = 6,611,724$

7. Sum of squared cell totals/sample size (pond $\times$ protein, two-way table) $= (2,284^2 + 2,532^2 + \ldots + 1,037^2 + 1,249^2) = 6,045,354$

8. Sum of squared cell totals/sample size (protein $\times$ vitamin C, two-way table) $= (1,290^2 + 1,012^2 + \ldots + 846^2 + 711^2) = 6,041,024$

9. If there were replications, sums of squared for each replicate totals would have to be computed. But in this case, there are no replications; therefore, there is no within the group variation.

10. Correction factor (C) $= (\text{Grand total})^2/n = 15,567^2/(3 \times 3 \times 5) = 5,385,440$

11. SS of Pond $= 3 - C = 6,006,526 - 5,385,440 = 621,086$

12. SS of vitamin C $= 4 - C = 5,949,866 - 5,385,440 = 564,426$

13. SS of Protein $= 5 - C = 5,423,355 - 5,385,440 = 37,915$

14. $SS_{Pond \times Vit\,C}$ (interaction) $= 6 - 3 - 4 + C = 6,611,724 - 6,006,526 - 5,949,866 + 5,385,440 = 40,772$

15. $SS_{Pond \times Protein}$ (interaction) $= 7 - 3 - 5 + C = 6,045,354 - 6,006,526 - 5,423,355 + 5,385,440 = 914$

16. $SS_{Protein \times Vit\,C}$ (interaction) $= 8 - 4 - 5 + C = 6,041,024 - 5,949,866 - 5,423,355 + 5,385,440 = 53,243$

17. $SS_{Pond \times Protein \times Vit\,C}$ (interaction) $= 2 - (6 + 7 + 8) + (3 + 4 + 5) - C = 6,714,259 - (6,611,724 + 6,045,354 + 6,041,024 + (6,006,526 + 5,949,866 + 5,423,355) - 5,385,440 = 10,464$

Now, these parameters are summarized in an ANOVA table (Table 7.34) in which MS values are obtained from SS dividing by their respective df, and F-values are obtained by dividing the MS values by $SS_{pond-vit\,C-protein}$.

Table 7.34 ANOVA table for factorial design.

Sources	SS	df	MS	F	Inference
C	5,385,440				
SS_{Pond}	621,086	2	310,543	475	***
SS_{vitC}	564,426	4	141,107	216	***
$SS_{Protein}$	37,914	2	18,957	29	***
$SS_{-Pond-vitC}$	40,772	8	5,096	7.79	***
$SS_{-Pond-Protein}$	914	4	228	0.35	ns
$SS_{-Protein-vitC}$	53,243	8	6,655	10.18	***
$SS_{-Pond-vitC-Protein}$	10,464	16	654		
	Total df	44		$F_{0.01,(8,16)}$	$= 6.19$

Table 7.35 Multiple comparisons (post-hoc tests) based on each factor.

Factor 1: Pond		N	Subset 1	Subset 2	Subset 3
Tukey's HSD(*a,b*)	3.00	15	209.67		
	2.00	15		331.73	
	1.00	15			496.47
	Sig.		1.000	1.000	1.000

Factor 2: Protein		N	Subset 1	Subset 2	Subset 3
Tukey's HSD(*a,b*)	20.00	15	308.07		
	30.00	15		351.33	
	40.00	15			378.47
	Sig.		1.000	1.000	1.000

Factor 3: Vitamin C		N	Subset 1	Subset 2	Subset 3	Subset 4
Tukey's HSD(*a,b*)	5.00	9	224.89			
	4.00	9		288.67		
	3.00	9		300.44		
	2.00	9			363.78	
	1.00	9				552.00
	Sig.		1.000	.861	1.000	1.000

Once the ANOVA is completed, multiple comparisons can be done among the treatments within each factor. However, if a factor has three or more than three graded levels, regression analysis can be done for trend analysis, as described in Section 8.2. Results of multiple comparisons are presented (Table 7.35) for each factor as in Section 7.4.

Multiple comparisons show that all three ponds differ significantly ($P < 0.05$). The first pond produced the largest fish followed by the second pond, and the third pond produced the smallest fish. Similarly, feed with the third level of protein (40%) gave the highest followed by the second (30%), and the lowest level of protein (20%) produced the smallest fish. However, all of the levels of vitamin C differ, except the third and fourth. The first level of vitamin C gave the highest followed by second, and the level 5 produced the smallest fish.

Most the researchers have problems with interpreting data when two factors have interactions. Use of graphs would make this clearer. For example, instead of parallel lines, the final weights of fish are shown closer to higher levels (lower mg·kg^{-1} diet) of vitamin C tested in three different ponds (Figure 7.16). A similar trend can be seen between protein and vitamin C levels (Figure 7.17).

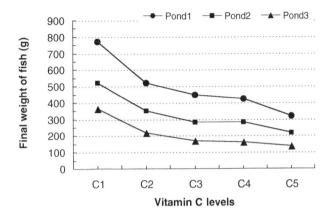

Figure 7.16 Interaction effects of Pond and vitamin C levels.

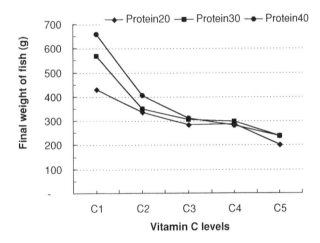

Figure 7.17 Interaction effects of protein and vitamin C levels.

7.6   Questions

1. In what way are experimental designs important?
2. Which design is more suitable for an experiment in a pond system and why?
3. What are the differences between Latin square and factorial designs?
4. If you have a limited number of experimental units, how do you replicate your treatments?

7.7   Practical exercises

Ex. 1. Table 7.36 shows the final mean weights (g) of fingerlings of different fish species after nursing for 2 weeks in polyculture in four ponds. Analyze the

Table 7.36 Final mean weights (g) of fingerlings.

Pond No.	Tilapia	Grass Carp	Rohu
1	15.1	12.0	10.2
2	14.5	12.3	11.1
3	14.3	11.2	10.1
4	14.1	11.5	10.3

following data using the appropriate statistical tools, present the results in graphical form, and write the results and discussion.

Ex. 2. Data in Table 7.37 of fish productivity ($t^{-1} \cdot ha^{-1} \cdot year^{-1}$) were obtained from an experiment in which four types of culture systems were compared. Write possible null hypotheses, analyze the data using appropriate statistical tools, and present the data using superscripts to show the comparisons among the means.

Ex. 3. Scores shown in Table 7.38 were given by the three members of a taste panel for the taste of meat from different species of fish. Test whether there are any differences among the six species of fish. Present results in tabular and graphical forms (using lines and superscripts) and write conclusions.

Ex. 4. A fertilization trial was conducted to evaluate the effects of chicken manure (100 and 200 $kg^{-1} \cdot ha^{-1} \cdot week^{-1}$) and urea (0 and 30 $kg^{-1} \cdot ha^{-1} \cdot week^{-1}$) on the growth of tilapia in ponds. Treatments were randomly allocated to 16 ponds assumed to be uniform in every aspect. Final mean weights (g) of fish are recorded (Table 7.39) after 150 days. Analyze the data and present the results (table and graph) and conclusions.

Table 7.37 Productivity of fish in polyculture ($t^{-1} \cdot ha^{-1} \cdot year^{-1}$).

Replication	Carp Monoculture	Tilapia Monoculture	Carp + Tilapia	Tilapia + Catfish
1	1.4	2.5	4.0	5.0
2	1.2	3.0	4.5	4.5
3	1.5	2.8	3.5	4.4
4	1.6	3.0	3.0	6.0

Table 7.38 Taste scores of different fish species.

Panel Members	Fish Species					
	Snakehead	Catfish	Grass Carp	Rohu	Tilapia	Sea Bass
1	5	1	3	2	4	6
2	6	2	3	1	5	4
3	6	3	2	1	4	5

Table 7.39 Mean individual weights of tilapia (g) obtained from the fertilization trial.

	Treatments			
	100 kg CM ha^{-1}·week^{-1}		200 kg CM ha^{-1}·week^{-1}	
Replication	0 kg urea ha^{-1}·week^{-1}	60 kg urea ha^{-1}·week^{-1}	0 kg urea ha^{-1}·week^{-1}	60 kg urea ha^{-1}·week^{-1}
1	150.6	225.1	173.8	320.4
2	128.9	210.6	161.3	301.9
3	137.2	218.5	155.6	296.4
4	130.5	224.3	166.4	306.7

Chapter 8

Testing and exploring relationships

8.1 Background

Chapter 7 covered basic methods of designing experiments and analyzing the results in which only one or a few factors and few levels of these factors were included. However, as mentioned earlier, in an actual biological system, an attribute or variable can be influenced by many factors and their levels simultaneously. These factors may vary themselves in different situations or locations and affect the attributes of others differently. Although factorial design and multifactor ANOVA help us investigate the relationships or interactions among the factors, it assumes that effects are linear, which means they increase or decrease at a constant rate. But in biological systems, there can be three or more fixed or independent factors, which produce a range of values with certain trends. The values may follow increasing or decreasing trends at varying rates. In addition, when a factor is acting upon one variable, there might be several other factors affecting the same variable; and, at the same time, these factors might affect many variables that might interact with each other. Some of these factors can be controlled and tested, whereas others are impossible to control but their effects can be separated by considering them as blocks. There can be even more factors that are completely out of control but still vary with time, season, or any other factor; these are called covariates.

In aquaculture production systems, a number of factors could bring about desired changes. Identifying these factors, exploring the nature of these relationships, and measuring them could benefit the people who depend on this. Therefore, in this chapter, attempts have been made to describe methods of hypothesis testing on relationships, which include proper designing and conducting of experimental or survey research and use of appropriate statistical methods for analysis (Figure 8.1). This should serve as one step further in explaining or solving the problems of complexity of the real world or nature.

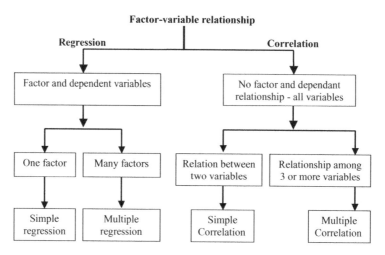

Figure 8.1 Types of factor-variable relationship.

8.2   Single-factor regression

Whether there is any cause and effect or dependency of one factor (effect) on the other (cause factor) is confirmed and/or measured by regression analysis. The factor (cause) brings a significant change (effect) on certain characteristics, often called dependent or response variables. For example, an increase in the level of nitrogen fertilizer in fish ponds increases fish yield, increased feeding rate increases growth rate, and so on. While designing survey or experimental research, it is very important for the researchers to identify the factor and the response variable. The levels of the factor should be sufficiently low as well as high so that an actual complete relationship can be established. Although the minimum numbers of level is three, more is better. In some cases, due to a lack of sufficient experimental units, researchers use fewer levels. There is some confusion about whether a level should be replicated. For regression analysis, it is better to increase the level of the factor rather than have replication for the levels. For example, if a researcher has six ponds for a trial, it is better to have six treatment levels, e.g. 0, 1, 2, 3, 4, and 5 kg $N\cdot ha^{-1}\cdot day^{-1}$, than to have three treatment levels, such as 0, 3, and 6 kg $N\cdot ha^{-1}\cdot day^{-1}$, with two replications.

8.2.1   Simple linear regression

Exploring relationships starts with a simple regression analysis, in which only one factor is considered to affect the variable, keeping or assuming the other potential factors constant. Preliminary regression analysis starts with drawing a scattered diagram to see the nature of data points followed by deciding the appropriate analysis. By looking at the distribution of data points in Figure 8.2 (left), anyone can guess that there is a linear relationship between fish production

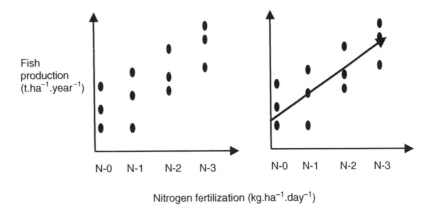

Figure 8.2 Scatter diagram (left) and a line drawn to represent data points (right).

and level of nitrogen where cause (level of nitrogen) and effect (fish production) are very clear. If a factor brings a constant increase or decrease in a variable, the relationship is linear; therefore, linear regression analysis has to be performed. In linear regression analysis, attempts are made to find a straight line, which represents the trend that the data points depict. The best fitted line is drawn through the points in such a way that the sum of squared deviations to the points from the line is minimal. Therefore, it is often called the least square deviations (LSD) method. Once the straight line is found, its slope or gradient is calculated, computing the change in response variable (ΔY) per unit change in factor (ΔX), i.e. $b = \Delta Y / \Delta X$ (Figure 8.3).

When the trend or slope (b) is positive, it means that there is an increase in response variable with the increase in factor. When the slope is negative, it means the response variable declines when the factor increases (Figure 8.4).

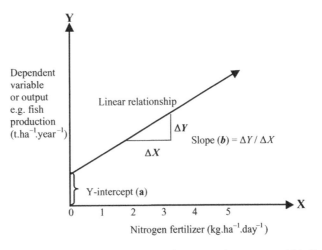

Figure 8.3 Slope of the line as relationship between factor X and response variable Y.

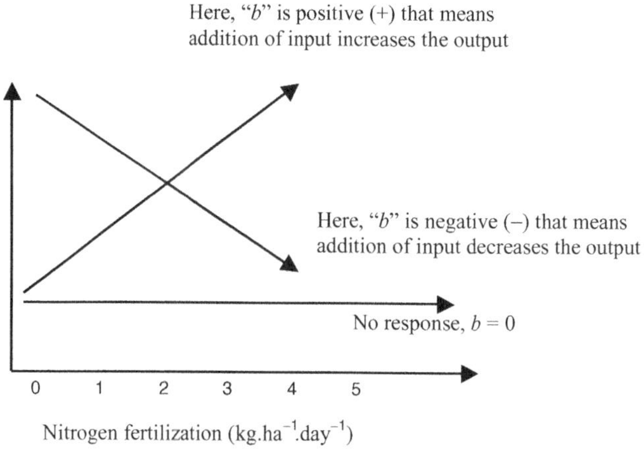

Here, "*b*" is positive (+) that means addition of input increases the output

Here, "*b*" is negative (−) that means addition of input decreases the output

No response, $b = 0$

Nitrogen fertilization (kg.ha^{-1}.day^{-1})

Figure 8.4 Relationship between factor and variable can be either positive (+) or negative (−) depending upon the situation.

The simple linear relationship is expressed mathematically as shown below, the coefficients of which are shown in Figure 8.3.

$$Y = a + bX$$

Where,

Y is a response variable that depends on factor X

X is a fixed or independent factor that affects variable Y

a is the constant or intercept, i.e. the level of Y at 0 level of X.

b is the slope of line or change in Y per unit change in factor X, i.e. $b = \Delta Y / \Delta X$

To describe simple linear regression, an example is given here. A simple data set is given in Table 8.1 in which weights of fish from 0 to 12th week were found in a trial. Table 8.2 shows the detailed calculations for the working equation. Here,

Age of fish (wks) is an independent factor "X"

Weight of fish (g) is dependent factor "Y"

Total age (ΣX) = 78, $n = 13$, and mean age ($\bar{X}$) = 6 wks

Total weight (ΣY) = 568, $n = 13$, and mean weights ($\bar{Y}$) = 43.7 g

The linear model: $Y = a + b \times \bar{X}$

Where regression coefficient (b) is computed as:

$$b_{yx} = \Sigma[(X_i - \bar{X})(Y - \bar{Y})] / \Sigma(X_i - \bar{X})^2$$

Table 8.1 Weekly mean weights (g) of fish reared in a pond.

Age (week)	0	1	2	3	4	5	6	7	8	9	10	11	12
Weight (g)	10.3	16.5	22.2	27.6	33.2	38.5	42.3	48.6	55.6	62.1	65.3	70.5	75.3

Table 8.2 Rearrangement of table for the computation of slope (b).

Age (wks)	Weight (g)	$/x - \bar{x}/$	$/y - \bar{y}/$	$(x - \bar{x})(y - \bar{y})$	$(x - \bar{x})^2$	$(y - \bar{y})^2$
0	10.3	−6	−33	200	36	1,115
1	16.5	−5	−27	136	25	739
2	22.2	−4	−21	86	16	462
3	27.6	−3	−16	48	9	259
4	33.2	−2	−10	21	4	110
5	38.5	−1	−5	5	1	27
6	42.3	0	−1	0	0	2
7	48.6	1	5	5	1	24
8	55.6	2	12	24	4	142
9	62.1	3	18	55	9	339
10	65.3	4	22	86	16	467
11	70.5	5	27	134	25	719
12	75.3	6	32	190	36	999
Total 78	568	0	0.0	990.8	182	5,404
Mean 6	43.7					

Here,

$$b_{yx} = \Sigma[(X_i - \bar{X})(Y - \bar{Y})]/\Sigma(X_i - \bar{X})^2$$
$$b_{yx} = 990.8/182 = 5.44$$

The linear model is $\bar{Y} = a + b \times \bar{X}$

Using values of b, $\bar{X}$, and $\bar{Y}$, constant (a) can be computed as:

$$a = \bar{Y} - b\bar{X} = 43.7 - (5.44 \times 6) = 11.03 \text{ g}$$

R-square value can be computed as:

$$r^2 = \left\{ \frac{\Sigma(Y - \bar{X})(Y - \bar{Y})}{\sqrt{\Sigma(X - \bar{X})^2 \times \Sigma(Y - \bar{Y})^2}} \right\} = \left\{ \frac{990.8}{\sqrt{182 \times 5404}} \right\}$$
$$= 0.98 \text{ or } 99.8\%$$

Here, the strength of the relationship between the weight of fish and the growth period (week) is 0.998, which means 99.8% variation is explained by the fitted straight line. Only 0.2% is the residual variance, which is explained by unknown factors.

Here,

Intercept or the constant (a) = 11.03 g
Slope or growth rate (b) = 5.44 g/wk
Linear model: $Y = a + b \bar{X}$

$$Y = 11.03 + 5.44X$$

Based on the results, it can be concluded that the estimated weight of fish at stocking (0 week age), i.e. intercept, $a = 11.03$ g, although observed data have been reported as 10.3 g. They grew by 5.44 g every week. Using this equation, prediction for any week, assuming that growth rate is constant, can be made. For example, what would be the weight of fish at the 20th week?

Here,

$$a = 11.03, \quad b = 5.44, \quad \text{and} \quad X = 20$$

Substituting these in the equation, we get

$$Y_{20wk} = 11.03 + (5.44 \times 20) = 119.8 \text{ g}$$

This shows that, at the end of the 20th week, fish will reach about 120 g.

8.2.2   Hypothesis testing

In regression, an attempt is made whether any trend or relationship between the factor and the response variable is tried, rather than comparing the output of two levels. Many researchers get confused about whether they should use factorial ANOVA followed by t-test or multiple range test or regression. Whenever possible, regression analysis should be performed rather than the use of factorial ANOVA. The well-fitted trend line or the model allows intrapolation, and every level of factor produces different outputs.

Before performing hypothesis testing, it is necessary to know about the degree of representation of data points or variability by the fitted line, i.e. r-square (r^2) value. It is also known as the coefficient of determination, which serves as an indicator of how well the data points are fitted in the model. It is measured on a scale of 0.0 to 1.0. When r^2 is 1.0, 100% variability is represented or explained by the model. In other words, the two variables have perfect association. If a model has an r^2 value of 0.75, it represents about three-fourths (75%) of the original variability. The remaining 25% is called residual variance, which means that there are other factors attributed to the remaining variation. Therefore, the higher the r^2 value, the better the model and the prediction made based on the model. In regression models or equations, coefficients such as a, b, c, etc. are used to signify their contributions to the variable. It is very important to know whether these coefficients are significant or not, and no matter whether they are small or large. This will be described with an example in the next section.

There are three main points while testing hypothesis for linear regression:

1. There is no significant relationship between the response variable and the factor, which means slope (b) = 0.
2. There is no significant difference between slopes of two lines, i.e. $b_1 = b_2$.
3. Intercept is not significant, which means constant (a) = 0.

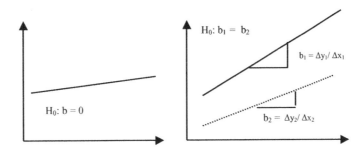

Figure 8.5 Comparison of slopes with x-axis (left) and between two slopes (right).

On the left of Figure 8.5, the slope of a line is not much different from 0, which means it is almost parallel to the x-axis. Therefore, it needs to be tested statistically against zero value. Similarly, on the right of Figure 8.5, the slopes (b_1 and b_2) of the two lines (dotted and solid) look different. In order to confirm whether they differ or not, they need to be tested. Their slopes can also be tested against zero value. To test these hypotheses for slopes, mainly two methods are used: ANOVA and Student's t-test. In this section, both of these are described using the data given in Tables 8.1 and 8.2, and a summary of the ANOVA table is shown in Table 8.3.

Here,

First, to test the null hypothesis (H_0): $b = 0$

Regression $SS = [\Sigma\,(X - \bar{X})(Y - \bar{Y})]^2/\Sigma\,(X - \bar{X})^2 = (990.8)^2/182 = 5{,}394$

Total $SS = \Sigma\,(Y - \bar{Y})^2 = 5{,}404$

Error or residual $SS = $ Total SS – Reg. $SS = 5{,}404 - 5{,}394 = 10$

$F_{0.05,1,11} = 4.84$, $P < 001$, reject H_0, which means the relationship between the dependent variable and the factor is highly significant. The slopes can also be tested using Student's t-test, similar to the testing of two treatment means:

$$t = (b - 0)/SE_b$$

From the previous example: $n = 13$, $b = 5.44$ g/wk

$$SE_b = \sqrt{\text{Residual MS}/(\Sigma(X - \bar{X})^2}$$
$$t = (b - 0)/SE_b = (5.44 - 0)/\sqrt{(0.89/182)} = 77.8^{**}$$

Table 8.3 ANOVA table for regression.

Source of Variation	SS	df	MS	F
Total	5,404	12	450	
Regression	5,394	1	5,394	6,068.2
Error or residual	10	11	0.89	

Here, $t_{0.05,11} = 2.201$. Therefore, reject H_0, which means the slope of the line is significantly higher than zero (0). In other words, the weekly growth rate of fish (5.44 g) is significantly higher than zero (0). Therefore, the fish in question show a significant growth every week, i.e. at the rate of 5.44 g.

Similarly, two slopes can be compared as in the case of two treatment means using Student's t-test with the null hypothesis (H_0): $b_1 = b_2$, as shown below:

$$t_{0.05 \cdot n-2} = (b_1 - b_2)/SE_{b_1 - b_2}$$

As in the multiple comparisons for means, it is possible to test multiple regression slopes, which is expressed as

$$(H_0) : b_1 = b_2 = b_3 = \ldots = b_n$$

In a similar way, coefficients of each factor obtained from nonlinear regression analysis can also be tested against hypothetic values (e.g. 0) and can also be compared between two of their corresponding coefficients.

8.2.3   Nonlinear regression

In well-controlled production systems, e.g. laboratories or factories, a linear relationship continues between the factor and the response variable, which means that every unit of additional input adds the same amount of output indiscriminately, regardless of the scale. However, in most biological systems, there are other factors acting at the same time which can't be controlled, but they can limit the production or the response variables. Therefore, the linear relationship no longer remains, and nonlinear regression analysis is necessary to explain such types of relationships. For example, when we increase the level of nitrogen fertilizer to 3 kg·ha^{-1}·day^{-1} or more, then the production per unit may not increase at the same rate afterward. Further addition of the fertilizer even might reduce the production due to toxicity or excessive growth of plankton, which causes oxygen depletion. This type of relationship is called a quadratic relationship. In Figure 8.6, it can be clearly seen that the extended straight line does not represent the points for 4 kg·ha^{-1}·day^{-1}, but the curved (quadratic) line looks better representative or better fitted with the points.

In many cases, an additional amount of nitrogen fertilizer used at the lower levels might give higher additional production than its previous unit, i.e. increasing at increasing rate. This type of relationship is called an exponential relationship. This is quite common in growth at younger stages of fish and other animals. Figure 8.7 presents three of the most common types of relationships between a response variable and a factor.

Quadratic and exponential are the most common nonlinear relationships in the biological system, including aquaculture production system. Although there are other types of relationships, e.g. cubic, growth, logarithmic, etc., they are not as common; therefore, these relationships are not covered in this book.

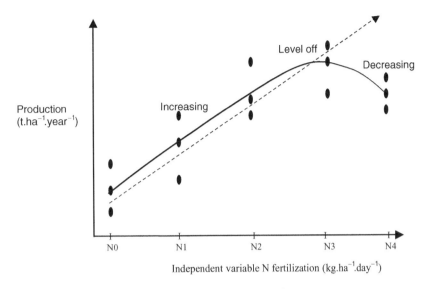

Figure 8.6 Relationship between nitrogen fertilizer and fish production.

As in the case of linear relationship, other types of relationships can also be expressed in mathematical equations or models, viz.:

Quadratic:

$$Y = b_0 + b_1 + b_2 X^2$$

Exponential:

$$Y = a \times e^{bx}$$

Where,

Y is the response variable that depends on factor X

X is the fixed or independent factor that affects variable Y

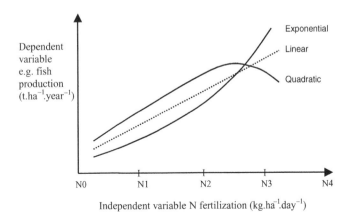

Figure 8.7 Types of relationship or trend that a data set of data may follow.

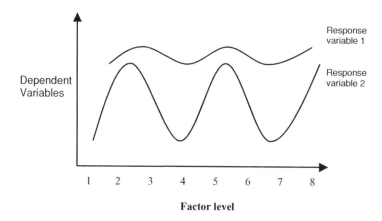

Figure 8.8 Polynomial relationships.

e is the natural logarithm
a is the constant or intercept, i.e. the level of Y at 0 level of X
b is the slope of the line or the change in Y per unit change in factor X
c is the coefficient of the second-degree term of factor X

Figure 8.8 shows the model of polynomial relationship (cubic and so on); however, this is rarely found in the biological system. If this exists, it can also be expressed in mathematical term by adding terms in the quadratic model:

$$Y = b_0 + b_1 X + b_2 X^2 + b_3 X^3$$
$$Y = b_0 + b_1 X + b_2 X^2 + b_3 X^3 + b_4 X^4$$

Where, d and f are coefficients of the third- and fourth-degree terms of factor X.

8.2.4   Model formulation and selection

In many cases, it may not be clear whether linear, quadratic, or any other types of line can best represent all of the data points. Preliminary model selection starts by drawing a scattered plot of the data generated or collected by the researcher. By looking at the points, their distribution, and trend, an experienced person can judge or guess the type of relationship that suits them best. However, regression analysis can assist in choosing the best model based on certain criteria or computed values, such as significance of the model or probability and the strength of the relationship (r^2). Therefore, in order to choose the best model, a scatter plot is needed to see the trend and outliers. The outliers are those data points that are far away from most of the others, which need special attention as a single outlier can change the trend line. They can be discarded with adequate justification to improve the model. However, without any reasonable justification,

Table 8.4 Summary of main regression models with CM as an independent variable.

Dep.	Models	Rsq.	df	F-value	Sigf.	b_0	b_1	b_2	b_3
NFY	LIN	0.654	25	47.30	0.000	2,321.18	5.0595		
NFY	QUA	**0.914**	24	127.92	0.000	1,029.14	18.0385	−0.0135	
NFY	CUB	0.919	23	86.45	0.000	7,742.82	22.1247	−0.0241	6.9E-06
NFY	EXP	0.652	25	46.75	0.000	2,207.95	0.0013		
NFY	LOG	0.832	25	123.73	0.000	−4,179.0	1582.98		

they shouldn't be discarded. Once potential models are developed, selection of the best model is complete. The first criterion is that the model should be significant. If more than one model is significant, then selection is done based on the r^2 value. If r^2 values are very close, select the simplest model, i.e. linear, for easier explanation and understanding. The determinants or the coefficients, i.e. a, b, c, etc., are tested for their statistical significance, meaningful or higher than zero. Hypothesis can also be tested whether there are two lines and whether their slopes are significantly different. Prediction for the values of Y can be made for given values of X once the model is valid or significant. The example below explains the selection steps.

Table 8.4 is the summary of outputs of regression analysis using one of the statistical packages in which the relationship between chicken manure (CM), i.e. X, and net fish yield (NFY), i.e. Y, was analyzed.

Here,

Dependent variable names appear in the first column as "Dep.," e.g. NFY.

Types of model appear in the second column as "Models," e.g. linear quadratic, cubic, exponential, and logarithmic.

r^2 values for the models appear in the third column named "Rsq."

df is in the fourth column

The F−value appears in the fifth and probability (P) values appear under the "Sigf" column.

Intercept (a) values appeared in the seventh column, followed by the coefficients of X values, respectively, in other columns.

Based on the results, mathematical models can be formulated as follows:

Linear: $Y = 2321.18 + 5.06X$ $(n = 27, P = 0.00, r^2 = 0.654)$

Quadratic: $Y = 1029.14 + 18.04X − 0.0135X^2$ $(n = 27, P = 0.00, r^2 = 0.914)$

Cubic: $Y = 774.282 + 22.1247X − 0.0241X^2 + 0.0000069X^3$ $(n = 27, P = 0.00, r^2 = 0.919)$

Exponential: $Y = ae^{bx} = 2207.95e^{0.0013X}$

Or

$L_n Y = L_n (2207.95) + 0.0013X$ $(n = 27, P = 0.00, r^2 = 0.652)$

Logarithmic: $Y = a + b \operatorname{Log}_X$ $(n = 27, P = 0.00, r^2 = 0.832)$

$Y = −4179.0 + 1582.98 \operatorname{Log}_X$

Selection of the appropriate model is a very important part of nonlinear regression analysis. The following steps should be followed:

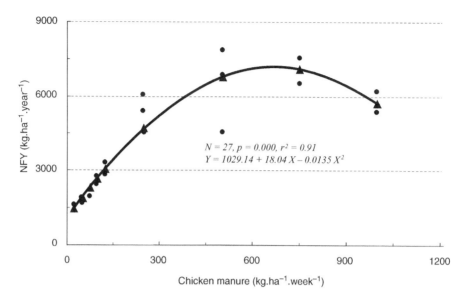

Figure 8.9 Quadratic relationship between chicken manure and net fish yield.

Step I: Select significant models only (i.e. F Sigf or $P < 0.05$).
Step II: If there is more than one significant model, select the one with higher r^2.
Step III: If r^2 values are quite close, select the simplest model, which is easier to describe or justify based on the constant and the trend line, e.g. linear should be preferred to quadratic and exponential, and they all should be preferred to the cubic model.
Step IV: Once a model is selected, results should be presented by plotting a graph along with its equation, number of samples taken, and the coefficient of determination (r^2).

In this particular example, all types of models obtained from a single set of data and presented in Table 8.4 are significant ($P = 0.000$). Now we have to determine which model represents or explains the highest variability. In other words, the model which has higher r^2 should be selected. For example, the cubic model has the highest, i.e. 0.919. That means about 92% of the variability is explained by the factor (chicken manure). Similarly, the quadratic model has 0.914 (91%), which is very close to the r^2 of the cubic model. In this case, the quadratic model is simpler and makes more sense than the cubic model. A cubic relationship is quite difficult to explain and normally is not applicable in most biological systems. Therefore, the quadratic model should be selected (Figure 8.9). ANOVA of regression (Tables 8.5 and 8.6) shows the coefficients of the model and their significance.

Results of the regression are presented below:
Dependent variable: NFY Method: Quadratic
Multiple R   0.95616
R^2 0.91424

Table 8.5 ANOVA table for quadratic regression.

Source of Variation	df	SS	MS
Regression	2	106,346,711.2	53,173,355.6
Residuals	24	9,976,007.3	415,667.0

F = 127.92297; Signif. F = 0.0000.

Table 8.6 Variables in the equation.

Variable	B	SE B	Beta	T	Sig T
CM	18.038453	1.566838	2.883755	11.513	0.0000
CM**2	−0.013453	0.001577	−2.136642	−8.530	0.0000
(Constant)	1029.144461	229.359372	4.487	0.0002	

Adjusted R^2   0.90709
Standard Error   644.72240

The selected quadratic model can be mathematically expressed as:

$$Y = 1029.14 + 18.04X - 0.0135X^2$$

Where, $n = 27$, $P = 0.000$, $r^2 = 0.91$.

The model shows that it is highly significant with high coefficient of determination as 91% of the variability is explained by the model. Only 9% is residual variance, which can be due to other unknown factors. Statistical outputs are not shown in the main text of any report or publication, but they should either be in an appendix or kept with the researcher so that they can be checked or traced back, if necessary. Conclusion from the analysis should be summarized as:

- About 1.0 ton of fish can be produced without chicken manure.
- About 18 kg of fish·ha^{-1}·year^{-1}can be increased ($P < 0.05$) by adding 1 kg·ha^{-1}·week^{-1} = 52 kg·ha^{-1}·year^{-1} chicken manure up to 600 kg·ha^{-1}·week^{-1}.
- Use of excess chicken manure (>600 kg·ha^{-1}·week^{-1}) reduces the fish yield.
- Maximum production level $(x) = -b/2a$, i.e. $18.04/(2 \times -0.0135) = 668$ kg.

Regression analysis can also be used for interpolation or extrapolation. Predicting or estimating a value of response variable (Y) for a level factor/input within the range is used to determine the relationship. For example, finding the value for X_2 as shown in Figure 8.10, where data used for determining the trend were between X_1 and X_4, is called interpolation, whereas predicting a value for a level of factor/input outside the range used to determine the relationship is called extrapolation, e.g. Y value for X_5 as shown in Figure 8.10. Normally, it is dangerous and should not be done except for the specific purpose of prediction.

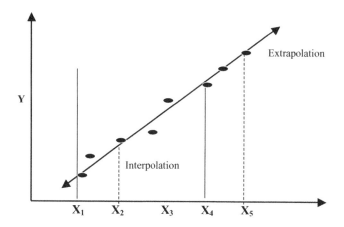

Figure 8.10 Interpolation and extrapolation using the best fitted model.

8.3   Multiple regression

In reality, response variables are affected by many independent factors simultaneously. Therefore, any study of impacts of a factor in isolation may not represent the true picture of the actual phenomenon, unless there is a well-controlled environment. Especially if a researcher is carrying out a trial in outdoor conditions and even collecting data from a survey of a wide range of environments, multiple regression analysis is necessary. For example, fish growth is affected by pond fertilization (N, P, K, etc.), feeding rate, temperature, DO, and several other factors. In a biological system, the relationships between these independent factors and dependent variables are not necessarily linear. However, as nonlinear relationships are quite complicated to deal with, they are beyond the scope of this book. In this section, methods of linear multiple regression are described with an example.

Multiple linear regression model is represented by the equation:

$$Y = a + b_1 \times X_1 + b_2 \times X_2 + \ldots + b_n \times X_n$$

8.3.1   Methods

Multiple regression identifies a model initially and iteration is carried out repeatedly, altering the model by adding or removing a predictor variable based on "stepping criteria." Iteration or the search for a new predictor is terminated when stepping is no longer possible with the stepping criteria, or when a specified maximum number of steps has been reached. In the resulting model, if ANOVA shows significance, that means at least one factor has significant effect, but it does not indicate which factors have significant effects; therefore, we must consult the table for coefficients for each factor. The best-fitted or most appropriate model is the one that includes all of the factors whose coefficients are significant.

There are main two methods of multiple regression analysis:

1. Forward selection or stepwise method
2. Backward elimination method

The forward selection method selects the most important variables serially. Therefore, it is possible to identify or rank variables based on their importance as it quickly determines the most important variable first, followed by the others serially. For example, if there are six variables from $X1$ to $X6$, the forward selection method would show the following results:

Model 1: $Y = a + b_2 X_2$
Model 2: $Y = a + b_2 X_2 + b_1 X_1$
Model 3: $Y = a + b_2 X_2 + b_1 X_1 + b_5 X_5$

Variables X_3, X_4, and X_6 were discarded because their coefficients were not significant, i.e. $P > 0.05$. The final selected model is Model 3, as it includes all of the significant variables.

The backward selection method starts by including all of the variables at first, then iterates by discarding insignificant variables step-by-step and keeps only significant ones at the final model. This method quickly identifies the least important factors easily. For example, with the six variables, X_1 to X_6, the backward selection method shows the following results:

Model 1: $Y = a + b_2 X_2 + b_1 X_1 + b_5 X_5 + b_3 X_3 + b_4 X_4 + b_6 X_6$
Model 2: $Y = a + b_2 X_2 + b_1 X_1 + b_5 X_5 + b_4 X_4 + b_3 X_3$
Model 3: $Y = a + b_2 X_2 + b_1 X_1 + b_5 X_5 + b_4 X_4$
Model 4: $Y = a + b_2 X_2 + b_1 X_1 + b_5 X_5$

As in the forward selection method, the variables X_3, X_4, and X_6 are discarded because they are insignificant ($P > 0.05$). The final model chosen is Model 4.

8.3.2   Example

For the purpose of demonstrating the method, an example of air pollution is described here, hoping that aquaculture researchers will also be able to analyze some environmental data.

Table 8.7 shows the data on air pollution in 20 selected American cities (Sokal and Rohlf 1969). The dependent variable (Y) recorded is the annual arithmetic mean concentration of sulfur dioxide ($\mu g \cdot m^{-3}$) as an indicator of air pollution. Among the six predictor variables, two are human-related, and the remaining are ecological variables which are as follows:

Y – SO_2 in air ($\mu g \cdot m^{-3}$)
X_1 – temperature (°F)
X_2 – no. of enterprises (>20 workers)
X_3 – population ('000)
X_4 – wind speed ($m \cdot hr^{-1}$)
X_5 – precipitation/rainfall (inch)
X_6 – no. of rainy days·year^{-1}

Table 8.7 Human and ecological variables of 20 American cities (Sokal and Rohlf 1969).

Name of Cities	Y	X1	X2	X3	X4	X5	X6
Little Rock	13	61.0	91	132	8.2	48.52	100
San Francisco	12	56.7	453	716	8.7	20.66	67
Denver	17	51.9	454	515	9.0	12.95	86
Hartford	56	49.1	412	158	9.0	43.37	127
Washington	29	57.3	434	757	9.3	38.89	111
Miami	10	75.5	207	335	9.0	59.8	128
Atlanta	24	61.5	368	497	9.1	48.34	115
Chicago	110	50.6	3,344	3,369	10.4	34.44	122
New Orleans	9	68.3	204	361	8.4	56.77	113
Baltimore	47	55.0	625	905	9.6	41.31	111
Detroit	35	49.9	1,064	1,513	10.1	30.96	129
Kansas City	14	54.5	381	507	10.0	37.00	99
Buffalo	11	47.1	391	463	12.4	36.11	166
Cincinnati	23	54.0	462	453	7.1	39.04	132
Cleveland	65	49.7	1,007	751	10.9	34.99	155
Columbus	26	51.5	266	540	8.6	37.01	134
Philadelphia	69	54.6	1,692	1,950	9.6	39.93	115
Pittsburgh	61	50.4	347	520	9.4	36.22	147
Houston	10	68.9	721	1,233	10.8	48.19	103
Seattle	29	51.1	379	531	9.4	38.79	164

Analyzing the data, Y as the dependent variable and $X_1 - X_6$ as independent factors or predictors in multiple regression, the results obtained are presented in Tables 8.8 and 8.9 using forward and backward selection methods, respectively.

$$Y - SO_2 \text{ in air } (\mu g \cdot m^{-3})$$

Factors: X_1, X_2, X_3, X_4, X_5, and X_6

Table 8.8 Results from stepwise or forward selection method.

		Coefficients[a]								
		Unstandardized		**Standardized Coefficients**			**Correlations**			**Collinearity Statistics**	
Model	B	Std. Error	Beta	t	Sig.	Zero-order	Partial	Part	Tolerance	VIF
1 (Constant)	13.721	4.930		2.783	.012					
X2	2.974E-02	.005	.811	5.871	.000	.811	.811	.811	1.000	1.000
2 (Constant)	71.665	27.025		2.652	.017					
X2	2.692E-02	.005	.734	5.620	.000	.811	.806	.706	.927	1.079
X1	−1.002	.461	−.284	−2.174	.044	−.483	−.466	−.273	.927	1.079
3 (Constant)	83.963	25.087		3.347	.004					
X2	2.715E-02	.004	.740	6.265	.000	.811	.843	.712	.926	1.080
X1	−1.823	.561	−.516	−3.248	.005	−.483	−.630	−.369	.512	1.955
X5	.854	.391	.341	2.185	.044	−.163	.479	.248	.529	1.889

[a] Dependent Variable: *Y*

Table 8.9 Results from backward selection method.

	Coefficients[a]										
	Unstandardized Coefficients		Standardized Coefficients			Correlations			Collinearity Statistics		
Model	B	Std. Error	Beta	t	Sig.	Zero-order	Partial	Part	Tolerance	VIF	
1 (Constant)	101.324	54.754		1.851	.087						
X1	−1.653	.885	−.471	−1.868	.084	−.477	−.460	−.221	.220	4.544	
X2	4.564E-02	.019	1.244	2.404	.032	.811	.555	.284	.052	19.151	
X3	−1.71E-02	.019	−.477	−.918	.375	.713	−.247	−.109	.052	19.290	
X4	−2.117	3.215	−.090	−.659	.522	.210	−.180	−.078	.741	1.349	
X5	.753	.583	.302	1.292	.219	−.174	.337	.153	.256	3.899	
X6	−1.05E-02	.200	−.010	−.052	.959	.267	−.015	−.006	.401	2.495	
2 (Constant)	99.569	41.712		2.387	.032						
X1	−1.622	.626	−.462	−2.590	.021	−.477	−.569	−.295	.408	2.450	
X2	4.560E-02	.018	1.243	2.495	.026	.811	.555	.284	.052	19.110	
X3	−1.70E-02	.018	−.475	−.952	.357	.713	−.247	−.108	.052	19.152	
X4	−2.170	2.944	−.093	−.737	.473	.210	−.193	−.084	.821	1.217	
X5	.733	.410	.293	1.787	.096	−.174	.431	.204	.482	2.073	
3 (Constant)	75.460	25.487		2.961	.010						
X1	−1.522	.602	−.433	−2.528	.023	−.477	−.547	−.284	.428	2.334	
X2	4.760E-02	.018	1.297	2.675	.017	.811	.568	.300	.054	18.686	
X3	−1.99E-02	.017	−.557	−1.164	.263	.713	−.288	−.131	.055	18.190	
X5	.711	.403	.285	1.766	.098	−.174	.415	.198	.485	2.063	
4 (Constant)	83.616	24.775		3.375	.004						
X1	−1.810	.555	−.516	−3.264	.005	−.477	−.632	−.370	.516	1.939	
X2	2.749E-02	.004	.749	6.374	.000	.811	.847	.723	.931	1.074	
X5	.844	.390	.338	2.164	.046	−.174	.476	.246	.528	1.895	

[a] Dependent Variable: Y

8.3.3   Interpretation of results

Table 8.10 shows a summary of three models obtained from the forward selection method. The first Model selects the most important factor, i.e. X_2, Model 2 adds the next important and significant factor, i.e. X_1, and Model 3 again adds X_5 as it is also significant; then it stops, as there are no other significant factors in the factor list. Model 3 is the one to select as it includes all of the possible significant factors and excludes all the factors that have no significant effect. The r^2 value increases with the increase in number of significant factors. Similarly, it increases as the number of insignificant variables are excluded from the model.

It does not matter which method we use as both of the methods ultimately give the same results, e.g. the third model obtained from the forward selection method (Table 8.8) and the fourth model from the backward selection method (Table 8.9). Both methods stop and produce the final model when all of the factors that have significant effects are included.

Table 8.10 R^2 values of the models from the forward selection method.

		Model Summary[d]		
Model	R	R Square	Adjusted R Square	Std. Error of the Estimate
1	.811[a]	.657	.638	16.0967
2	.855[b]	.732	.700	14.6510
3	.891[c]	.793	.755	13.2537

[a] Predictors: (Constant), $X2$
[b] Predictors: (Constant), $X2$, $X1$
[c] Predictors: (Constant), $X2$, $X1$, $X5$
[d] Dependent Variable: Y

The model is expressed as using the coefficients appearing in a column under "B" in Tables 8.8 or 8.9 as:

$$Y = 83.963 - 1.823 X_1 + 0.02715 X_2 + 0.854 X_5$$

Where, $n = 20$, $P = 0.000$, $r^2 = 0.793$
The description of the model is summarized as follows:

- Per unit increase in temperature (X_1) decreases 1.823 μg $SO_2 \cdot m^{-3}$ as there is negative partial correlation.
- Increase of 1 enterprise (X_2) can increase 0.0275 μg $SO_2 \cdot m^{-3}$.
- Increase of 1 inch of rainfall per year can increase 0.854 μg $SO_2 \cdot m^{-3}$.
- Based on the model, predictions can be made (described in Section 8.3.4).

8.3.4 Prediction

What would be the minimum and maximum SO_2 levels in a city where annual temperature ranges from 45 to 75°F, if there are 2,000 enterprises and the average annual precipitation is 50 inches?

Solution:
For minimum temperature: 45°F

$$Y = 83.963 - 1.823 X_1 + 0.02715 X_2 + 0.854 X_5$$
$$= 83.963 - 1.823(\times 45) + 0.02715(\times 2000) + 0.854(\times 50)$$
$$= 99 \ \mu g \ SO_2 \cdot m^{-3}$$

For maximum temperature: 75°F

$$Y = 83.963 - 1.823 X_1 + 0.02715 X_2 + 0.854 X_5$$
$$= 83.963 - 1.823(\times 75) + 0.02715 \times (2000) + 0.854(\times 50)$$
$$= 44 \ \mu g \ SO_2 \cdot m^{-3}$$

The range is 44−99 μg $SO_2 \cdot m^{-3}$

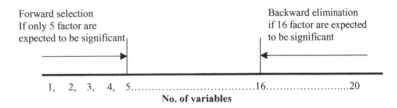

Figure 8.11 Description of selection of model in multiple regression.

8.3.5   Selection of method

It is a question whether a researcher should use the forward selection or backward elimination method. The basic principle is that, if we have many variables in the list but expect only a few variables to have significant effects, then we would use forward selection. However, on the other hand, if we expect many variables to have significant effects and only a few will be discarded, then the backward elimination method is suitable. For example, if there are 20 variables/factors and we think that only 5 factors will have effects, then it's better to go from the front (Figure 8.11). However, if we think 16 factors have significant effects, or only 4 factors will needed to be discarded, then start from the back and use the backward selection method. This is very similar to selecting a path of low distance to reach your destination faster.

8.4   Correlation and parametric test

Correlation is a measure of association between or among variables. The basic assumption of correlation is that there is no dependency of one variable on the other. The two variables go together, caused by one or more other factors showing some sort of association. The degree of association is expressed as the correlation coefficient (r). Correlation can be either positive or negative. Positive correlation means that an increase in one variable is paired with the increase in the other, e.g. length and weight of fish. Whereas in the case of negative correlation, increased value of the first variable is coupled with decreased value of the second variable. Fish survival and daily weight gain normally have negative correlation because high survival means more fish remained in the system and share limited space and food, which results in smaller fish, i.e. low daily weight gain.

The correlation coefficient (r) can be computed by using the following formula:

$$r = \frac{\text{Covariance}}{\sigma_X \cdot \sigma_Y}$$

$$r = \frac{\Sigma(X - \bar{X})(\Sigma Y - \bar{Y})}{\sqrt{[N\Sigma X^2 - (\Sigma X)^2][N\Sigma Y^2 - \{\Sigma Y\}^2]}}$$

As with other statistical parameters, the correlation coefficient can also be computed using a working formula without means, such as:

$$r = \frac{N\Sigma\,XY - (\Sigma\,X)(\Sigma\,Y)}{\sqrt{[N\Sigma\,X^2 - (\Sigma\,X)^2][N\Sigma\,Y^2 - \{\Sigma\,Y\}^2]}}$$

Where,

N is the number of data pairs

$\Sigma\,XY$ is the sum of the products of data pairs

$\Sigma\,X$ is the sum of the first variable, i.e. X

$\Sigma\,Y$ is the sum of the second variable, i.e. Y

$\Sigma\,X^2$ is the sum of the squared observations of variable X

$\Sigma\,Y^2$ is the sum of the squared observations of variable Y

The correlation coefficient is measured within ± 1 scale or $-1 \geq r \geq +1$. If two variables have a correlation value of $+1$, that means they have perfect (100%) positive correlation; similarly, if they have a correlation value of -1, that means perfect (100%) negative correlation. Similarly, correlation of zero (0) means the two variables have no association at all. However, in reality these conditions do not occur.

The squared correlation coefficient, i.e. r^2, represents the proportion of common variation between two variables; therefore, it is also called the coefficient of determination (see also Section 8.2.3). It is normally expressed in percentage, e.g. 92%, which means 92% of the variation is due to their association and the remaining 8% is due to other factors or conditions, also called residuals.

For the sake of understanding this in qualitative terms, correlation (r) has been crudely categorized as high (0.8–1.0), medium (0.6–0.8), fair (0.4–0.6), and low (<0.4); however, the reliability of correlation has to be tested for its significance, which depends mainly on the size of sample. In linear correlation, it is assumed that the residuals are distributed normally. If the size of sample is at least 50, data are most likely normal, and if it reaches 100 or over, then a researcher need not worry about meeting normality assumptions.

As in regression, correlation coefficients can also be tested as to whether they are significantly higher than zero (0) or compared between themselves. The significance of correlation coefficients depends on their magnitudes as well as the size of the samples. Although a low correlation coefficient from a large sample size can be significant, reliability or the significance of the correlation coefficient increases with its magnitude. A difference of 0.10 between two correlation coefficients may not be significant if the two coefficients are 0.10 and 0.20, but the same degree of difference, i.e. 0.10, can be highly significant when the two coefficients are 0.80 and 0.90 in the same data sets.

For the purpose of describing the method, an example is given. From a trial of length (L) and weight (W), measurements of 29 male tilapias were recorded simultaneously (Table 8.11) to see the correlation. For this, the cross-products of L and W and their squares need to be calculated as shown in Table 8.11.

Table 8.11 Total length (L) and net weight (W) of 29 male tilapia.

SN	Length (L, cm)	Weight (W, g)	LW	L^2	W^2
1	21.0	163	3,423.0	441.0	26,569
2	16.4	74	1,213.6	269.0	5,476
3	14.4	51	734.4	207.4	2,601
4	22.5	173	3,892.5	506.3	29,929
5	21.5	143	3,074.5	462.3	20,449
6	18.5	118	2,183.0	342.3	13,924
7	19.8	120	2,376.0	392.0	14,400
8	19.8	137	2,712.6	392.0	18,769
9	22.5	184	4,140.0	506.3	33,856
10	18.5	112	2,072.0	342.3	12,544
11	20.5	145	2,972.5	420.3	21,025
12	21.5	191	4,106.5	462.3	36,481
13	20.0	115	2,300.0	400.0	13,225
14	17.4	86	1,496.4	302.8	7,396
15	20.0	140	2,800.0	400.0	19,600
16	17.0	89	1,513.0	289.0	7,921
17	19.5	120	2,340.0	380.3	14,400
18	23.4	223	5,218.2	547.6	49,729
19	19.8	138	2,732.4	392.0	19,044
20	18.5	107	1,979.5	342.3	11,449
21	21.5	160	3,440.0	462.3	25,600
22	19.4	108	2,095.2	376.4	11,664
23	19.0	125	2,375.0	361.0	15,625
24	23.0	175	4,025.0	529.0	30,625
25	18.2	103	1,874.6	331.2	10,609
26	20.5	134	2,747.0	420.3	17,956
27	18.2	108	1,965.6	331.2	11,664
28	17.5	101	1,767.5	306.3	10,201
29	16.5	84	1,386.0	272.3	7,056
Sum	566.3	3,727	74,956.0	11,186.9	519,787

Here,

$N = 29$

$\Sigma LW = 74,956.0$

$\Sigma L = 566.3$

$\Sigma W = 3,727$

$\Sigma L^2 = 11,186.9$

$\Sigma W^2 = 519,787$

Therefore, correlation coefficient (r):

$$r = \frac{29(74,956.0) - (566.3 \times 3,727)}{\sqrt{[\{29(11,186.9 - (566.3)^2)\}\{29(519,787 - (3,727)^2)\}]}} = 0.95$$

This shows that length and weight of tilapia had high correlation (0.95), or in other words, it can be explained that there is 95% chance of increasing weight when there is an increase in length or vice versa.

Table 8.12 Correlation between *X*1 and *X*5 factors shown in Table 8.7.

		X1	X5
	Correlations		
X1	Pearson Correlation	1	.686**
	Sig. (2-tailed)	.	.001
	N	20	20
X5	Pearson Correlation	.686**	1
	Sig. (2-tailed)	.001	.
	N	20	20

** Correlation is significant at the 0.01 level

Most of the statistical packages designed for correlation analysis also show whether the correlation coefficient is significant, as shown in Table 8.12; e.g. taking X_1 and X_5 factors used in multiple regression (Table 8.7). The table shows that variables X_1 and X_5 have highly significant ($P < 0.01$) medium positive correlation ($r = 0.686$).

8.5   Nonparametric tests for regression and correlation

Relationship between two variables can also be tested nonparametrically. Two tests are particularly popular. They are Spearman's rank correlation and Kendall's coefficient of concordance. Both of them are described briefly with examples.

8.5.1   Spearman's rank correlation

This is also called the nonparametric bivariate correlation (also regression) method, in which Spearman's rank correlation coefficient (r_s) is computed by using the following equation:

$$r_s = 1 - [(6\Sigma d^2)/(n^3 - n)]$$

For example, data sets of viability (%) of tilapia eggs and ammonia nitrogen are shown in Table 8.13. The data are ranked separately, then differences in ranks of these corresponding values of the two variables are computed. The differences are squared to obtain $\Sigma\ d^2$, then the coefficient of Spearman's rank correlation is computed, which is compared with the value in Appendix A10.
Here,
Spearman's Rank correlation coefficients (r_s)
$= 1 - 6\Sigma\ d^2/(n^3 - n)$
$= 1 - 6\times(533)/(12^3 - 12)$
$= -0.864$

Table 8.13 Calculation of Sspearman's correlation.

Wks	1	2	3	4	5	6	7	8	9	10	11	12	Total
Data:													
Viability (%)	74.9	77.6	80.2	81.7	78.2	85.3	81.1	85.6	83.0	86.4	85.3	87.2	
NH_3-N (ppm)	38	38.8	37.1	38.2	34.5	33.2	32.1	31.9	32.4	31.9	30.5	29.3	
Ranks:													
Viability (R)	1	2	4	6	3	8.5	5	10	7	11	8.5	12	
NH_3 (ranks)	12	11	9	10	8	7	5	3.5	6	3.5	2	1	
Rank "d"	−11	−9	−5	−4	−5	1.5	0	6.5	1	7.5	6.5	11	
d^2	121.0	81.0	25.0	16.0	25.0	2.3	0.0	42.3	1.0	56.3	42.3	121.0	533.0

From the table in Appendix A10, $r_{s0.05,12} = 0.587$

Reject H_0; therefore, the conclusion is that there is significant negative correlation (as the coefficient is negative) between the variables of egg viability and the ammonia nitrogen of the water of breeding ponds.

8.5.2　Kendall's rank correlation or Kendall's coefficient of concordance

This is a nonparametric method for a data set that has three or more variables to be tested. It is also called the nonparametric multivariate correlation method. An example from one of our trials is shown in Table 8.14 for the purpose of describing this method.

In this example, three variables were measured from the same groups of fish which were fed with varying feeding rates. Feeding has obvious effects on the three variables, such as weight of the female, gonadosomatic index (GSI), and the seed output. As one of the data, such as GSI, was percent data, nonparametric test was performed.

Here,

Null hypothesis (H_0): There is no association among the three variables.

$$M = 3, n = 18, \Sigma R = 171, \Sigma R^2 = 513.0$$

Kendall's coefficient of concordance (W) can be computed as:

$$W = [\Sigma R^2 - (\Sigma R^2/n)]/[M^2(n^3 - n)/12]$$
$$= [513.0 - (171/18)]/[3^2(18^3 - 18)/12] = -0.2549$$

This shows that the combined association among the three variables, as indicated by Kendall's coefficient of concordance, is negative. However, it needs to be determined whether it is significant (higher than zero or not) by converting it

Table 8.14 Association among mean weight of female tilapia (g), mean GSI, and average egg output (g) per spawning.

Data			Ranks			
Mean Weight of Female (g)	GSI (%)	Egg (g)	Weight	GSI	Egg	Sum
72.4	3.2	17.5	6	9	2	17.0
68.7	2.6	20.8	5	4	3	12.0
64.3	3.4	35.6	3	11	7	21.0
114.9	4.3	74.1	8	14.5	16	38.5
154.0	2.7	45.7	12	5	8	25.0
136.6	3.8	49.3	11	13	10	34.0
172.7	2.8	30.3	16	6.5	6	28.5
157.3	3.3	29.9	13	10	5	28.0
170.4	3.0	48.1	15	8	9	32.0
68.5	3.7	53.7	4	12	11	27.0
58.8	6.1	13.1	2	18	1	21.0
57.8	4.6	29.8	1	17	4	22.0
129.7	2.5	73.5	10	3	15	28.0
113.8	4.3	70.8	7	14.5	14	35.5
122.1	4.5	55.3	9	16	12	37.0
202.9	2.8	65.9	18	6.5	13	37.5
159.2	2.3	149.6	14	1.5	18	33.5
200.2	2.3	83.7	17	1.5	17	35.5
					171.0	513.0

into a χ^2 value as shown below:

$$\chi^2 = W \times M(n-1)$$
$$= (-0.2549) \times 3(18-1) = -13.0$$

$\chi^2_{0.05,17} = 27.587$ (from χ^2 table in Appendix A2). As the computed value is lower than the value in the table, we have to accept H_0, which means that there is no significant ($P > 0.05$) association among the three variables.

However, researchers should be careful because there can also be interactions among the three variables. Two of them might have positive association, whereas the other two might have negative association due to which the combined coefficient is sometimes confusing and difficult to explain. In this case, it is better to check pairwise correlations between possible pairs of two variables. This can be easily performed by choosing suitable tools, available in most statistical packages. For an example, results are shown in Table 8.15 for which data from Table 8.14 were used.

The results showed that weight of female and the GSI have highly significant negative association ($P = 0.01$), but egg output and the GSI have no significant association ($P > 0.05$). Similarly, weight of female and egg output have no significant association ($P > 0.05$). The combined association among all of these variables might have been affected by their potential positive and negative

Table 8.15 Statistical test for correlation coefficients using non-parametric methods.

			Weight	GSI	Egg
Kendall's tau_b	Weight	Correlation Coefficient	1.000	−0.449(**)	0.307
		Sig. (2-tailed)	–	0.010	0.075
		N	18	18	18
	GSI	Correlation Coefficient	−0.449(**)	1.000	−0.211
		Sig. (2-tailed)	0.010	–	0.224
		N	18	18	18
	Egg	Correlation Coefficient	0.307	−0.211	1.000
		Sig. (2-tailed)	0.075	0.224	–
		N	18	18	18

* Correlation is significant at the 0.05 level (2-tailed) – but there is none in this particular case.
* Correlation is significant at the 0.01 level (2-tailed).

(as seen in their coefficients) associations in Table 8.15, possibly resulting in nonsignificant combined association.

8.6   Multiple correlations

In real biological systems, there are a number of variables associated with others. The number of variables to be considered is completely dependent on the situation. It is not easy to fix which variables are needed to select for the collection of data. It is itself a research and basically guided by the objectives of the research. A researcher needs to identify those variables and explore and determine correlation among them. While designing any research and selecting the variables for relationship, researchers should be very careful. Some of the variables are intermediate in nature, which might be affected by other variables or factors. These variables might even be at different levels one after another caused by the same factor or many others. Wherever possible, instead of dealing with the intermediate variables, attempts should be made to identify the actual causal factor(s). Correlation analysis is only to see the association between two variables which might be affected by the same factors or different ones. Once the factors that are causing changes in one or many variables are identified, regression analysis is needed to determine the rate of change with per unit change in factor so that recommendation can be made based on the results. Length and weight of fish can vary simultaneously, caused by the amount or rate of feeding. In this case, it is clear that length and weight of fish are variables that have high correlation, and the rate or amount of feeding is a factor. However, fish weight and length may show positive correlation with intermediate variables, such as DO levels and plankton growth, and negative correlation with ammonia and nitrite levels. Correlation or regression between two variables does not fully explain the phenomenon. For instance, correlation or regression between fish survival and ammonia level, without taking fertilization rate and phytoplankton growth into account, is not complete research as it does not explain the actual causes. This

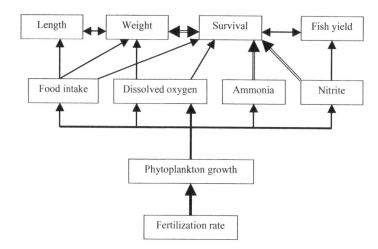

Figure 8.12 Schematic diagram to represent the complexity of factors in fertilized pond system.

would be considered partial work. Therefore, multiple correlation is extremely helpful in this case as they can explain the actual phenomenon of fish growth and survival, thereby yield, covering most variables involved.

Computing multiple correlation coefficients can be very time-consuming and complex (Figure 8.12), but it can be performed easily by using various statistical

Table 8.16 Partial correlation coefficients and their significance.

			Factors Controlling for Y			
X1	**X2**	**X3**	**X4**	**X5**	**X6**	
X1	1.0000	.2500	.2729	−0.1677	0.6968	−0.2953
	(0)	(17)	(17)	(17)	(17)	(17)
	$P = .$	$P = 0.302$	$P = 0.258$	$P = 0.493$	$P = 0.001$	$P = 0.220$
X2	0.2500	1.0000	0.9456	0.2759	−0.1219	−0.3298
	(17)	(0)	(17)	(17)	(17)	(17)
	$P = 0.302$	$P = .$	$P = 0.000$	$P = 0.253$	$P = 0.619$	$P = 0.168$
X3	0.2729	0.9456	1.0000	0.2957	−0.1140	−0.3524
	(17)	(17)	(0)	(17)	(17)	(17)
	$P = 0.258$	$P = 0 .000$	$P = .$	$P = 0.219$	$P = 0.642$	$P = 0.139$
X4	−0.1677	0 .2759	0.2957	1.0000	−0.1416	0.2209
	(17)	(17)	(17)	(0)	(17)	(17)
	$P = 0.493$	$P = 0.253$	$P = 0.219$	$P = .$	$P = 0.563$	$P = 0.363$
X5	0.6968	−.1219	−.1140	−.1416	1.0000	.2681
	(17)	(17)	(17)	(17)	(0)	(17)
	$P = 0.001$	$P = .619$	$P = .642$	$P = .563$	$P = .$	$P = .267$
X6	−0.2953	−0.3298	−0.3524	0.2209	0.2681	1.0000
	(17)	(17)	(17)	(17)	(17)	(0)
	$P = 0.220$	$P = 0.168$	$P = 0.139$	$P = 0.363$	$P = 0.267$	$P = .$

(Coefficient / (df) / 2-tailed Significance)
" . " is printed if a coefficient cannot be computed.

packages. In presence of other variables together at the same time, the correlation of any two variables can be estimated separately assuming/keeping other variables constant; this is called partial correlation. Table 8.16 shows that there are six variables and partial correlations between two variables, which can be seen from crosstab analysis as shown in Table 8.15.

8.7   Questions

Q1. What are the differences between regression and correlation?
Q2. Why is quadratic regression more important in biological research?
Q3. What are the applications of multiple regression?
Q4. In what ways are correlation analysis important?

8.8   Practical exercises

Ex. 1. After conducting an experiment and analyzing the data, suppose you have the following relationship: $Y = 1029 + 18.04X - 0.0135X^2$ ($n = 27$, $P = 0.000$, $r^2 = 0.91$) between chicken manure (kg·ha^{-1}·wk^{-1}) and fish yield (kg·ha^{-1}·wk^{-1}). Answer the following questions asked by a farmer who is planning to construct a fish farm with a 2-ha pond area:

a. What sort of effect does chicken manure have on fish yield?
b. How much fish can he produce without using chicken manure?
c. What is the maximum fish production he could get if this relationship holds true?
d. If he has only 800 kg of chicken manure, how much fish can he produce?

Ex. 2. Data in Table 8.17 were collected from an experiment conducted to investigate the relationship between the rate of chicken manure (CM) and net fish yield (NFY). Test whether there is a linear, quadratic, or exponential relationship between them, select the best fitted model, and describe it.

Table 8.17 Batch weights of tilapia cultured in ponds.

Observation	CM (kg·ha^{-1}·year^{-1})	NFY (kg·ha^{-1}·year^{-1})	Obs.	CM (kg·ha^{-1}·year^{-1})	NFY (kg·ha^{-1}·year^{-1})
1	25	1,520	15	125	3,325
2	25	1,485	16	250	4,540
3	25	1,640	17	250	5,420
4	50	1,890	18	250	6,045
5	50	1,920	19	500	6,850
6	50	1,680	20	500	7,860
7	75	2,250	21	500	4,540
8	75	1,950	22	750	6,540
9	75	2,010	23	750	6,530
10	100	2,480	24	750	7,520
11	100	2,540	25	1,000	6,200
12	100	2,770	26	1,000	5,350
13	125	3,020	27	1,000	5,600
14	125	2,835	–	–	–

Chapter 9

Advanced topics

9.1   Cluster analysis

Cluster analysis is an exploratory data analysis for classification where each cluster represents a class in which its members show close relationships or similar characteristics. It is also considered an important tool of discovery because it contributes to the definition of a formal classification scheme, such as taxonomy for related animals or plants, and also suggests statistical models to describe populations or to indicate rules for assigning new classes. Therefore, the purpose of cluster analysis is to discover a system of organizing observations into groups where members of the same groups share common properties. The association among members within the same cluster is stronger than between members of different clusters.

Clustering is a basically method of combining similar objects in a separate group. Simple grouping is possible by a visual observation of the data and frequency polygon or scatter plot in many cases. But if it involves complex multivariate data, classification is not possible by simple observation. Computer software is needed for classifying the groups and assigning the values to them.

Cluster analysis starts with preparing a data matrix in which objects are arranged in rows and observations are in columns. A table of either relative similarities (proximities matrix) or differences is created between all objects, and objects are combined into groups based on this. This section describes the basic principle of clustering using a single variable only.

9.1.1   Univariate cluster analysis

Clustering can also be done based on a single set of observations, which is called univariate cluster analysis. For example, fish can be grouped into high-value to low-value fish based on their market prices using cluster analysis. Table 9.1 shows the current price of six main species in Thailand.

Table 9.1 Approximate market prices of most common fish species in Thailand.

Fish Species	Price per kg (US$)
1. Nile tilapia (NT)	0.59
2. Silver barb (SB)	0.88
3. Catfish (CF)	1.18
4. Snake head (SH)	2.65
5. Tiger prawn (TP)	4.41
6. Marble goby (MB)	10.29

Table 9.2 Proximities matrix of market prices of various fish species in Thailand.

	NT	SB	CF	SH	TP	MB
NT	0.00	0.29	0.59	2.06	3.82	9.70
SB	**0.29**	0.00	0.30	1.77	3.53	9.41
CF	**0.59**	**0.30**	0.00	1.47	3.23	9.11
SH	**2.06**	**1.77**	**1.47**	0.00	1.76	7.64
TP	**3.82**	**3.53**	**3.23**	**1.76**	0.00	5.88
MB	**9.70**	**9.41**	**9.11**	**7.64**	**5.88**	0.00

Note: NT, Nile tilapia; SB, silver barb; CF, catfish; SH, snake head; TP, tiger prawn; MB, marble goby.

The proximities matrices for these fish species are computed from the price differentials in absolute value. These are also called distances between two pairs. For example, for the cell meeting NT and SB = 0.88 − 0.59 = 0.29 in Table 9.2.

These distances are called Euclidean distances. Clustering can also be done using squares of those distances between all the possible pairs of observations. These are called squared Euclidean distances. For example, the same NT and SB = $(0.29)^2 = 0.08$. Squared distances give more clear differences, and use of squared distances is similar to other statistical tools, e.g. ANOVA and the least squares criterion. Table 9.3 is the complete proximities matrix based on the squared differences as the distance measure.

The next step is the preparation of dendogram based on these matrices. These proximities matrices show a symmetrical pattern if divided by a diagonal line

Table 9.3 Squared proximities matrix of market prices of various fish species in Thailand.

	NT	SB	CF	SH	TP	MB
Nile tilapia (NT)	0.00	0.08	0.35	4.24	14.59	94.09
Silver barb (SB)	0.08	0.00	0.09	3.13	12.46	88.55
Catfish (CF)	0.35	0.09	0.00	2.16	10.43	82.99
Snake head (SH)	4.24	3.13	2.16	0.00	3.10	58.37
Tiger prawn (TP)	14.59	12.46	10.43	3.10	0.00	34.57
Marble goby (MB	94.09	88.55	82.99	58.37	34.57	0.00

from top left to bottom right. As the numbers in row and column entries are the same on each half of the matrix, only one half can be used for preparation of the dendogram. For our purpose, Table 9.4 has been drawn from Table 9.2 for simplicity.

Using these distances, classification is done. A dendogram (Figure 9.1) prepared based on the above method uses average linkage between groups (rescaled distance cluster combine).

Various statistical packages have options using other distance measures, e.g. Cosine, Chebychev, Block, Minkowski, customized, and so on. Some of these options contain further options themselves.

As an example, a comprehensive classification of more fish species as shown in Table 9.5 using the hierarchical method based on their market prices (univariate cluster analysis) is described. Proximity matrices are given in Table 9.6, and a dendogram (Figure 9.2) has been prepared by using computer software.

Table 9.4 Proximities matrix of market prices of various fish species in Thailand (extracted from Table 9.2).

	NT	SB	CF	SH	TP	MB	Clusters
NT	0.0						
SB	0.3	0.0					**Cluster 1**
CF	0.6	0.3	0.0				NT, SB, CF
SH	2.1	1.8	1.5	0.0			
TP	3.8	3.5	3.2	1.8	0.0		
MB	9.7	9.4	9.1	7.6	5.9	0.0	

	NT, SB, CF	SH	TP	MB	
NT, SB, CF	0.0				
SH	1.5	0.0			**Cluster 2**
TP	3.2	1.8	0.0		NT,SB,CF+ SH
MB	9.1	7.6	5.9		

	Cl1-SH	TP	MB	
Cl1-SH	0.0			
TP	1.8	0.0		**Cluster 3**
MB	7.6	5.9	0.0	NT,SB,CF,SH + TP

	Cl1-SH-TP	MB	
Cl1-SH-TP	0.0		
MB	5.9	0.0	**Cluster 4**
			NT,SB,CF,SH,TP+MB

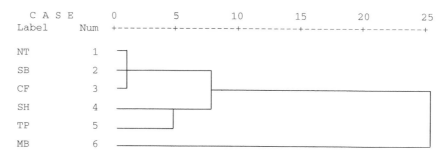

Figure 9.1 Classification of six main fish species based on their prices in Thailand.

Table 9.5 Market prices of various fish species in Thailand.

SN	Fish Species	Price (US$)
1	Carps	0.44
2	Nile tilapia	0.59
3	Silver barb	0.88
4	Catfish	1.18
5	Red tilapia	1.76
6	Giant gourami	2.21
7	Sea bass	2.35
8	Snake head	2.65
9	Crab	2.94
10	Tiger prawn	4.41
11	Giant prawn	7.35
12	Marble goby	10.29

Table 9.6 Proximity matrices.

	Squared Euclidean Distance											
Case	1: Carps	2: Nile tilapia	3: Silver barb	4: Catfish	5: Red tilapia	6: Giant gourami	7: Sea bass	8: Snake head	9: Crab	10: Tiger prawn	11: Giant prawn	12: Marble goby
1: Carps	0.00	0.02	0.19	0.55	1.74	3.13	3.65	4.88	6.25	15.76	47.75	97.02
2: Nile tilapia	0.02	0.00	0.08	0.35	1.37	2.62	3.10	4.24	5.52	14.59	45.70	94.09
3: Silver barb	0.19	0.08	0.00	0.09	0.77	1.77	2.16	3.13	4.24	12.46	41.86	88.55
4: Catfish	0.55	0.35	0.09	0.00	0.34	1.06	1.37	2.16	3.10	10.43	38.07	82.99
5: Red tilapia	1.74	1.37	0.77	0.34	0.00	0.20	0.35	0.79	1.39	7.02	31.25	72.76
6: Giant gourami	3.13	2.62	1.77	1.06	0.20	0.00	0.02	0.19	0.53	4.84	26.42	65.29
7: Sea bass	3.65	3.10	2.16	1.37	0.35	0.02	0.00	0.09	0.35	4.24	25.00	63.04
8: Snake head	4.88	4.24	3.13	2.16	0.79	0.19	0.09	0.00	0.08	3.10	22.09	58.37
9: Crab	6.25	5.52	4.24	3.10	1.39	0.53	0.35	0.08	0.00	2.16	19.45	54.02
10: Tiger prawn	15.76	14.59	12.46	10.43	7.02	4.84	4.24	3.10	2.16	0.00	8.64	34.57
11: Giant prawn	47.75	45.70	41.86	38.07	31.25	26.42	25.00	22.09	19.45	8.64	0.00	8.64
12: Marble goby	97.02	94.09	88.55	82.99	72.76	65.29	63.04	58.37	54.02	34.57	8.64	0.00

Note: this is a dissimilarity matrix.

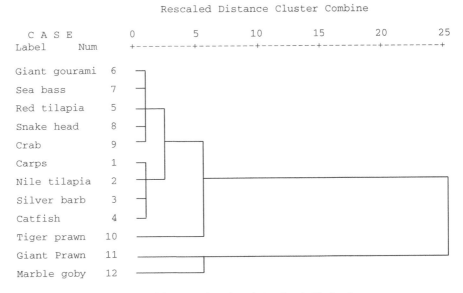

Figure 9.2 Classification of 12 fish species based on their prices in Thailand.

9.1.2   Multivariate cluster analysis

If clustering is done by taking into account many observations, it is called multivariate cluster analysis. The main point in multivariate cluster analysis is that a researcher has to decide which observations are relevant to include for analysis. For example, level of income or farm earning would be relevant for ranking the wealth of farmers, but age of the farmers may not be relevant. In this section, both types of cluster analysis are described.

In many cases, classification is done based on more than one observation. In such cases, squared distances are computed for each observation separately, then combined to prepare a table of final proximities matrices. That means squared distances of each cell for all the observations are added in the cells of final proximities matrices. This will be the basis for classification. However, this method is valid if the observations are in similar scales. But if scales are different, the observation with higher values can shadow the distances of the others, which have smaller values. In such cases, observations are transformed first, and then they are combined to standard scores before computing the separated distance matrices. There is difficulty in deciding whether to transform the data or not, because the choice of metric distance can result in vastly different proximities matrices. Researchers who need to use multivariate cluster analysis should research this further (see Bibliography and Webliography), as this is beyond the scope of this book.

9.2   Analysis of Covariance

Analysis of covariance (ANCOVA) is a more sophisticated method of ANOVA that further minimizes experimental error, removing effects of unavoidable factors or conditions other than the treatments that vary themselves and affect response variables. If the factors causing variabilities can't be controlled through experimental designs only, a statistical tool called analysis of covariance (ANCOVA) is used while analyzing the data for further partitioning of total variability attributable to those factors. ANCOVA is done by using concurrent variables, which are called covariates, with response or dependent variables.

In most biological research, it is quite difficult, sometimes not even possible, to find all the experimental units identical in all respects at the start of the trial, e.g. chemistry or nutrients in pond sediment and water, size of experimental animals, and so on. Measurement or analysis of pond water and bottom soil at the beginning of the experiment is important as effects due to the difference in these variables might be considerable but can be adjusted using ANCOVA (Schefler 1969). ANCOVA can also be useful in adjusting the values distracted by natural calamities or unavoidable circumstances, e.g. fish death due to unexpected diseases, floods, or others. It is necessary to test whether the significance of changes in covariates has resulted in any changes on the dependent variables, in other words, whether any significant correlation exists.

Similarly, while conducting a trial in outdoor conditions, there are several variables other than the ones considered in the design that may vary with time; these other factors are called changing covariates, e.g. water temperature, DO, pH, levels of ammonia, nitrite, and so on. Although these changing variables or factors cannot be controlled, especially in outdoor trials or in the real fields, they can be measured or recorded simultaneously at all the points when the dependent variables are measured repeatedly, e.g. daily, weekly, biweekly, monthly, and so on. ANCOVA helps us to determine whether these changing variables have any effects on dependent variable, such as fish growth over the experimental period. Therefore, ANCOVA is an analysis of variance on the residuals of the dependent variable after removing the influence of the covariate rather than on the original values themselves. ANCOVA is considered a combined method of ANOVA with regression. As in regression, covariance analysis determines whether any treatments show different responses with fluctuations of these changing variables. If there is any relation, covariance removes the effects of external factors by adjusting their influence on the dependent variables. The method of covariance analysis as described by Schefler (1969) has been applied to an example of data generated from an aquaculture research in this section.

Table 9.7 shows data on reproductive output (no. of eggs per female) of two groups of tilapia (normal and previously stunted), collected along with their mean weights. The objective was to determine whether stunting has any effects on seed output as compared with the normal group. It is obvious that weight of females has an impact on the seed output. Therefore, in order to separate the effect of female weight, a regression coefficient (b), i.e. regression of all the

Table 9.7 Egg output from normal and previously stunted tilapia.

| | | Egg Output (no./female) | |
	Female wt.	Original (*Y*)	Adjusted (*Y'*)
Normal tilapia (A)	72.4	18	32
	68.7	22	37
	64.3	37	54
	114.9	78	80
	154.0	48	39
	136.6	53	49
	172.7	31	17
	157.3	31	22
	170.4	50	37
Stunted tilapia (B)	68.5	59	74
	58.8	24	42
	57.8	28	46
	129.7	76	74
	113.8	71	74
	122.1	56	56
	202.9	70	48
	159.2	154	144
	200.2	85	64
Mean	123.6	55.0	55.0

weights (*Xs*) on all egg outputs (*Ys*), has been computed as described in Section 8.2 and was found to be 0.28. Original egg output data are adjusted using the following equation:

$$\text{Adjusted egg output } (Y') = /\text{Regression coefficient } (X - \text{grand mean}) - Y/$$
$$= /b(X_i - \overline{X}) - Y_i/$$

For example, for the first row data, adjusted seed output was (*Y'*) = (72.4 − 123.6)(0.28 − 18) = 32.

Egg output can be compared after adjusting the egg output data. The result obtained is the actual effect of treatment (stunting). For comparison purposes, *t*-test has been performed on the data (Table 9.8).

The *t*-test result showed that there was no significant ($P > 0.05$) difference in egg output between normal and stunted tilapia when using original data; however, after adjustment for covariance, egg output was found to be significantly different ($P < 0.05$), more due to stunting (Table 9.9). This is mainly because error variance is reduced after adjustment.

As the sizes of females in each group are different, data should be analyzed using covariance rather than simply *t*-test. Most researchers compare the size of female groups and the seed output using a simple t-test. For example, using *t*-test, we get the results as shown in Table 9.10.

Table 9.8 Two-sample t-test comparing egg output of Nile tilapia.

	Egg Output (Original Data)		Egg Output (Adjusted data)	
	Normal	Stunted	Normal	Stunted
Mean	40.90	69.14	40.93	69.11
Variance	339	1438	349	950
Observations	9	9	9	9
Pooled variance	888		650	
Hypothesized mean diff.	0		0	
df	16		16	
t stat	−2.010		−2.346	
$P(T \le t)$ one-tail	0.031		0.016	
t critical one-tail	1.746		1.746	
$P(T \le t)$ two-tail	0.062		0.032	
t critical two-tail	2.120		2.120	

Table 9.9 Detail of ANCOVA for the reproductive output (no. of eggs per female) of two groups of tilapia are given along with their mean weights.

	Normal Tilapia		Stunted Tilapia	
Group	Mean Female Weight (X_A)	Egg Output (Y_A)	Mean Female Weight (X_B)	Egg Output (Y_B)
1	72.4	18	68.5	59
2	68.7	22	58.8	24
3	64.3	37	57.8	28
4	114.9	78	129.7	76
5	154.0	48	113.8	71
6	136.6	53	122.1	56
7	172.7	31	202.9	70
8	157.3	31	159.2	154
9	170.4	50	200.2	85
Sum	1,111.2	368	1,113.0	622
Average	123.5	41	123.7	69

Table 9.10 shows that the sizes of the females are not different and the egg output is also not different ($P > 0.05$). As size of female has obvious relation with seed output, we need to analyze the data using covariance. For ANCOVA, all of the data are squared separately and factor and variable are multiplied to get the sum of crossproducts of each parameter, which is shown in Table 9.11. Here,

$$\Sigma XY = 47,519 + 87,477 = 134,996$$
$$\Sigma X = 1,111.2 + 1,113.0 = 2,224.2$$
$$\Sigma Y = 368 + 622 = 990$$
$$\Sigma X^2 = 153,256 + 162,754 = 316,010$$
$$\Sigma Y^2 = 17,764 + 54,524 = 72,288$$
$$\Sigma X_A = 1,111.2$$

Table 9.10 Results of the Student's *t*-test; comparison of weights.

	Female Weight			Egg Output	
	Normal	**Stunted**		**Normal**	**Stunted**
Mean	123.5	123.7		40.9	69.1
Variance	2007.3	3140.3		338.6	1437.8
Observations	9	9		9	9
df	16.000			16.000	
t stat	−0.008			−2.010	
$P(T \leq t)$ one-tail	0.497			0.031	
t critical one-tail	1.746			1.746	
$P(T \leq t)$ two-tail	**0.994**	*NS*		**0.062**	*NS*
t critical two-tail	2.120			2.120	

$\Sigma X_B = 1{,}113.0$
$\Sigma Y_A = 368$
$\Sigma Y_B = 622$
$N = 18$
$n_1 = 9$
$n_2 = 9$

Now, partitioning of sum of squares is done as follows:

1. For weight of female, i.e. variable X:

$$SS_{\text{Total}} = \Sigma X^2 - (\Sigma X)^2/n = 316{,}010 - (2{,}224.2)^2/18$$
$$= 41{,}181.7$$
$$SS_{\text{Weight}} = (\Sigma X_A)^2/n_1 + (\Sigma X_B)^2/n_2 - (\Sigma X)^2/N$$
$$= (1{,}111.2)^2/9 + (1{,}113.0)^2/9 - 2{,}224.2/18 = 0.2$$
$$SS_{\text{Within}} = SS_{\text{Total}} - SS_{\text{Weight}} = 41{,}181.7 - 0.17$$
$$= 41{,}181.5$$

Table 9.11 Data from Table 9.9 arranged for ANCOVA.

	X_A^2	Y_A^2	X_B^2	Y_B^2	$X_A Y_A$	$X_B Y_B$
	5,236	329	4,689	3,431	1,313	4,011
	4,725	487	3,455	576	1,517	1,411
	4,133	1,396	3,338	788	2,402	1,622
	13,193	6,023	16,834	5,745	8,914	9,834
	23,711	2,259	12,952	5,112	7,318	8,137
	18,658	2,804	14,919	3,100	7,234	6,801
	29,815	952	41,154	4,832	5,328	14,102
	24,736	968	25,345	23,670	4,894	24,493
	29,050	2,545	40,067	7,270	8,598	17,068
Sum	153,256	17,764	162,754	54,524	47,519	87,477

Table 9.12 Table for sum of squares.

Source	SS(X)	SS(Y)	SS(XY)
Total	41,181	17,800	12,624
Female weight	0	3588	25
Within	41,181	14,212	12,599

2. For egg output, i.e. variable Y:

$$SS_{Total} = \Sigma Y^2 - (\Sigma Y)^2/n = 72,288 - (990)^2/18 = 17,800$$
$$SS_{Weight} = (\Sigma Y_A)^2/n_1 + (\Sigma Y_B)^2/n_2 - (\Sigma Y)^2/N$$
$$= (368)^2/9 + (622)^2/9 - 990/18 = 3,588$$
$$SS_{Within} = SS_{Total} - SS_{Egg} = 17,800 - 3,588 = 14,212$$

3. Partition of SS for crossproducts (SSCP), i.e. XY:

$$SS_{Total} = \Sigma XY - (\Sigma X \times \Sigma Y)/N = 134,996 - (2224.2 \times 990)/9$$
$$= 12,624$$
$$SS_{Weight} = (\Sigma X_A \times \Sigma Y_A)/n_1 + (\Sigma X_B \times \Sigma Y_B)/n_2 - (\Sigma X \times \Sigma Y)/N$$
$$= (1,111.2 \times 368)/9 + (1,113 \times 622)/9 - (2224.2 \times 990)/18$$
$$= 25$$
$$SS_{Within} = SS_{Total} - SS_{Weight} = 12,624 - 25 = 12,599$$

4. Summary of SS partitioning is shown in Table 9.12.
5. Now regression coefficients, X on Y (b_1) and Y on X (b_2), can be computed as:
   i) $b_1 = \Sigma XY/\Sigma X^2 = 12,624/41,181.7 = 0.307$
   ii) $b_2 = \Sigma XY/\Sigma Y^2 = 12,599/41,181.5 = 0.306$
6. Covariance can be calculated as:

$$\Sigma(Y - Y_P)^2 = \Sigma(Y - Y_P)^2 - b_1 \Sigma XY = (17,800) - 0.307 \times 12,624$$
$$= 13,930$$
$$\Sigma(Y - Y_P)^2 = \Sigma(Y - Y_P)^2 - b_2 \Sigma XY = (14,212) - 0.306 \times 12,599$$
$$= 10,357$$

7. ANCOVA table is prepared as shown in Table 9.13.

In conclusion, although the t-test shows that there was no significant difference between the two groups of tilapia in terms of seed output, ANCOVA shows that the difference is significant. This is because the variation in weight has an impact on the seed output, which was masked when seed output was compared by using Student's t-test, neglecting the potential impact of weight of the fish. This clearly

Table 9.13 Final table of ANCOVA.

Source	df	SS	MS	F	Remarks
Female weight	1	3,573	3,573	5.175	* Significant at 5%
Within	15	10,357	690		
Within + Female wt.		13,930			

shows the importance ANCOVA. In fact, there can be other variables that might be associated with the egg output, e.g. gonad weight or GSI, nutritional conditions, water temperature, ammonia and DO, levels, and so on. The usefulness and the power of ANCOVA have not been used and realized by most aquaculture researchers because it is tedious and quite complex to understand and interpret. However, various computer software programs can perform ANCOVA very easily. It is therefore suggested to be familiar with the method as it would greatly strengthen the capacity of researchers on data analysis and reduce the chance of committing Type II error.

9.3   Multivariate ANOVA

As described in previous sections, two means are compared by using Student's *t*-test, taking only one dependent variable at a time. Similarly, ANOVA is used if many means are involved; however, it is only for a single dependent variable. In reality. several response variables are affected simultaneously by treatment factor(s). Instead of looking at just one response variable, measuring a few other relevant variables normally makes it easier to explain any given phenomenon. Table 9.14 briefly illustrates the type of test to be performed based on the number of variables involved.

Some of the variables might be the variables of intermediate steps of the whole process or system. All of the variables are associated with each other, i.e. change in one response variable might bring about positive or negative changes in other variables. All of the possible variables should be measured so that it is easier to explain and interpret the results. In other words, it is an attempt to understand the whole system rather than a part of it. In many cases, finding out the effects of only one response variable without knowing the effects on other variables

Table 9.14 Types of statistical tests based on the number of variables.

Groups	One dependent variable (univariate analysis)	Two or more dependent variables (multivariate)
Two groups	*t*-test	Hotelling's T^2
More than two groups	ANOVA and multiple comparison tests	MANOVA

means that the research actually remains incomplete. For an example, if a trial is designed to determine the effects of pond fertilization rate on fish yield per unit area, recording final weights of whole batches of fish would not be sufficient information for explanation, because yield is the final product of individual growth, which is affected by survival. All of these largely depend on the growth and availability of natural food organisms, which have direct relation with the levels of DO, ammonia, nitrite, and so on. All of these variables should be measured wherever and analyzed simultaneously so that actual direct effects of fertilization on these parameters can be estimated. Due to association among such variables, treatment effects can be masked easily. It is essential to separate the variances due to such associations. Therefore, multivariate analysis (MANOVA) is a must in such cases. It is possible that we may get significant effects of treatments from MANOVA, when we get marginally nonsignificant effects of the same factors obtained from univariate analysis (ANOVA), which means neither variable considered alone is significantly affected by the factor. Therefore, the MANOVA is more powerful than ANOVA. However, researchers should understand that some of the environmental parameters, such as water temperature (which doesn't depend on fertilization, but depends on other factors such as season and sunlight condition), should not be used as variables. Instead it should be used as covariate while analyzing the data, as described in Section 9.2.

As in the case of ANOVA, MANOVA may also show that a factor has significant effects on the variables but still does not point out which variables are particularly affected. Therefore, after multivariate test, if significant overall effect is shown, univariate ANOVA must be performed for individual variables to identify the variables that contribute to the significant overall effect.

MANOVA is also capable of determining whether two or more independent variables, and also their interactions, affect two or more normally distributed dependent variables, often called factorial MANOVA. As with factorial ANOVA, normality and homogeneity of variance are assumed. As in univariate factorial ANOVA, we can also test whether there are any interaction effects between/among the independent variables. As a preliminary step, univariate interactions can be analyzed before moving on to multivariate. However, results from MANOVA should be used for the final conclusion.

The multivariate model is expressed as:

$$Y = Xb + e$$

Where,
$Y = n \times p$
$X = n \times k$
$b = k \times p$
$e = n \times p$
p is dependent variables
k is parameters for each dependent variable
n is observations

For univariate ANOVA, *F*-value is calculated as the ratio of a treatment variance (mean square) to error variance, whereas for MANOVA, treatment effect is tested by calculating the ratio of the determinant (generalized variance) of the error SSCP matrix to the determinant of the sum of the treatment and error SSCP matrices. This ratio is called Wilks' Lambda (λ), i.e. error/(error + treatment). In other words, it is the proportion of variance in dependent variables that is not accounted for by the independent variable. Although Wilks' Lambda seems smaller compared with the F-value for ANOVA, it raises questions about the null hypothesis.

Table 9.15 Experimental design of the research.

Hapa Exchange Interval (days)	Feeding Level (% biomass)	
	Existing Level (0.75%)	High Level (1.5%)
5	5-E	5-H
15	15-E	15-H
60	60-E	60-H

MANOVA is described here in detail, using a data set from a trial conducted in a commercial hatchery to determine the effects of feeding levels and hapa exchange interval on the reproductive performance of Nile tilapia. The trial used a 3 × 2 factorial design, as shown in Table 9.15.

Six treatments (3 levels of hapa exchange intervals × 2 levels of feeding) were randomly allocated in 18 hapas (24 m × 5 m) installed in a single pond (size 5,300 m²). Mature broodfish were stocked at a density of 6 fish·m⁻² of hapa space with 1:1 sex ratio, which means 360 males and 360 females were stocked in each hapa. Numbers and batch weights (each replicate hapa) of males and females were recorded at stocking and on days 10, 25, 30, 50, 65, 80, 95, 110, and 120. Seed was harvested manually from the mouths of incubating females every 5 days and pooled to coincide the period of batch-weighing intervals. Number of seed was estimated based on the samples weights of 200 seed taken at each harvest. DO at 6 a.m. and temperature 3:30 p.m. inside each hapa were monitored a day before each batch weight measurement. Effects of hapa exchange and feeding levels on the reproductive outputs (weight of seed and the number) were assessed by using MANOVA.

In this trial, our main objective of the research, or the hatchery operator's main interest, was to determine the impacts of feeding rate and hapa exchange strategies (factors) on the number of seed as final output. As there are other intermediate variables associated with the seed output, e.g. seed weight, number and weight of fish, water temperature, and DO level, they were also collected and used for analysis.

The main hypotheses tested were:

- Increased feeding level increases size of broodfish, especially females, which helps increase seed output because larger females normally produce more eggs.

- Frequent replacement of fouled hapas with cleaned ones improves water exchange; therefore, it increases DO levels, which have positive impacts on fish growth and then reproductive performance.
- Frequency of hapa exchange increases the chance of broodfish escape; therefore, it has negative impacts on the fish remaining in the hapa, which ultimately has a negative effect on seed output.
- Water temperature and DO affect seed intermediate variables, thereby seed output, but these two factors were considered as covariates.

For statistical analysis purposes, data can be arranged serial-wise, i.e. hapa numbers 1–18 (shown in Appendix B1) and their corresponding data follow in other columns. This would be the same for most survey research, e.g. household numbers in the first column and their attributes in others. Outcomes of the analysis (Appendix B) and results are presented subsequently. The variable names are coded as shown below:

1. FTWT – total weight of female (kg)
2. FNO – total number of females per hapa
3. FSURV – survival of females (%)
4. FMWT – mean weight of individual females (g)
5. MTWT – total weight of males (kg)
6. MNO – total number of males per hapa
7. MSERV – survival of males (%)
8. MMWT – mean weight of individual males (g)
9. DENSITY – no. of fish per square meter of hapa space (no·m^{-2})
10. BIOMASS – total weight of males and females (kg·m^{-2})
11. SEEDNO – estimated number of seed at each harvest (×100,000)
12. SEEDWT – weights of seed harvested at each harvest (g)
13. DO6 – DO level measured at 6 a.m.
14. TEMP3 – water temperature measured at 3 p.m.

While analyzing the data, DO6 and TEMP3 were used as covariates, whereas all others were as dependent variables.

9.4 Interpretation of results

The most difficult and important part is to interpret and describe the results of the statistical analysis correctly and concisely in various presentation forms, e.g. written sentences, tables, and graphs. Almost all of the statistical software produce various types of graphs that can help interpret the results (e.g. Figure 9.3). However, the quality is not always very good; therefore, it is advised that graphical presentation be made using Microsoft® Excel or any other graphical software.

The General Linear Model (Appendix B4) lists all of the dependent variables used, their sum of squares, mean squares, *F*-values, Sig. (*P*), and partial eta

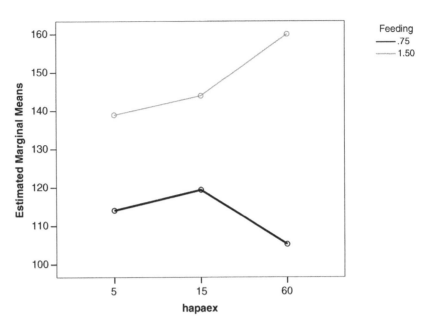

Estimated Marginal Means of FMWT

Figure 9.3 Significant interaction ($P = 0.048$) between feeding rate and hapa exchange (feeding $\times$ hapaex, Appendix B4) in mean weight of females.

squared values (effect size) for a model, intercept, and each of the variables. Sig. (P) is the main value that determines whether any variable is significant in the model, i.e. it should be less than 0.05 (or another level if chosen differently). For example, intercepts of SEEDNO and SEEDWT are significant because their corresponding P-values are 0.012 and 0.014; similarly, effects of feeding on these variables are also significant ($P = 0.02$ and 0.001, respectively), whereas effects of TEMP3 are nonsignificant on FNO and FSURV ($P = 0.504$ and 0.551). The strength of the relationship (r^2-values) appears at the bottom of the table, e.g. r^2-value for FTWT is 0.774 or 77.4% and for SEEDNO is 0.929 or 92.9%. In this way, significance of each variable and their strengths of relationship can be indicated.

Results of the multivariate corrected model (Appendix B4) can be summarized by saying that all of the variables used for analysis are affected significantly ($P < 0.05$) at least by one or more factors when the association effects among response variables were separated. Significant intercept in mean and total weights of females and males and seed outputs (seed weight and number) means that these values could be higher than zero even when they were without treatment factors, i.e. not fed and hapas were not exchanged.

The feeding rate didn't show any significant difference ($P = 0.330$ and $P = 0.209$) in seed outputs (SEEDNO and SEEDWT, respectively) when they were analyzed for univariate functions (Appendix B7) nor when compared using direct pairwise comparisons (Appendix B5). But the effect of feeding rate was significant ($P < 0.01$) when MANOVA was performed in the general linear model (Appendix B4), as mentioned earlier. This might cause confusion for the researchers about which results should be reported. Those who do not perform MANOVA will certainly report that there is no effect of feeding rate on the seed outputs (SEEDNO and SEEDWT), but that is not the case. After using MANOVA, it is found that feeding actually has an effect on seed outputs but indirectly. It affects other variables first, such as mean weights of fish, e.g. FMWT and MMWT which have ultimate effects on seed outputs. This shows the value of MANOVA.

MANOVA also allows for interaction effects e.g. in this example, feeding and hapa exchange show significant interaction effects on mean weights of females ($P = 0.048$, Appendix B4), which can also be shown in graphical form (Figures 9.3 and 9.4).

All four multivariate test statistics obtained from analysis are given in Appendices B6, B9, and B12. They are all significant ($P = 0.000$) for all the factors in this particular case. Tables from Appendices B13–B21 show the multiple range test performed together with MANOVA. Tukey's HSD shows that 60-day hapa exchange treatment has higher survival than 5 and 15 days (Appendix B19), whereas Duncan's test shows that all three treatments differ (as they appear in different columns), showing highest survival from 60-day exchange followed by 15-day and then 5-day intervals. As mentioned in Section 7.2.1 Tukey's test results should be selected if data are less normal.

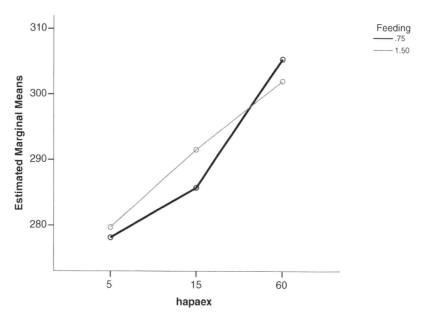

Figure 9.4 Number of female fish with the increase in hapa exchange interval for both the feeding levels (see Table 2.17).

9.5   Questions

Q1. Why is ANCOVA important?

Q2. What are the applications of cluster analysis?

Q3. MANOVA is the ultimate and highest level of statistical analysis. Explain briefly.

9.6   Practical exercises

Ex. 1. Data shown in Table 9.16 are the mean weights of common carp and grass carp (g) from a 16-week trial. They were raised in ponds (200 m^2) at 1 fish·m^{-2} stocking density and fed at 1%, 2%, and 3% biomass feeding rates per day. Analyze the data using the appropriate statistical tools.

Ex. 2. Data shown in Table 9.17 were from a trial with tilapia broodfish. The fish were fed with pellet feeds mixed with four different oils and the control feed. Muscle samples were collected and lipid analyzed. Using multivariate and covariance functions, analyze the data and write results.

Table 9.16 Mean weights (g) of common carp and grass carp.

Weeks	Common Carp			Grass Carp		
	1%	2%	3%	1%	2%	3%
0	36	38	42	33	33	36
1	38	41	46	35	36	39
2	40	45	50	36	39	43
3	41	48	54	38	42	48
4	43	52	59	40	45	53
5	45	55	64	41	49	59
6	47	60	69	43	52	65
7	49	64	75	45	57	72
8	50	69	82	46	61	79
9	52	74	89	48	66	88
10	54	80	97	50	71	97
11	56	86	105	51	77	107
12	58	93	114	53	83	119
13	59	100	124	55	90	131
14	61	107	135	56	97	145
15	63	115	147	58	105	160
16	65	124	160	60	113	177

Table 9.17 Lipid composition of muscle and water quality parameters.

Treatments		Lipid Groups in the Muscle (% of the total lipid)						Water Quality	
Oils	Fert.	SM	PC	PS	Cholesterol	TAG	Cholesters	Temp	DO at 6 a.m.
1	1	4.97	31.57	3.35	13.19	17.08	6.16	29	0.2
1	1	4.15	30.08	2.28	13.69	24.17	4.25	29.5	0.3
1	2	3.31	35.99	2.9	17.06	8.69	4.45	29.1	0.5
1	2	3.79	34.36	3.28	16.92	14.87	5.23	29.7	0.6
2	1	3.97	32.85	2.72	14.44	18.93	3.14	29.8	0.1
2	1	3.26	25.74	2.14	11.8	30.62	5.09	29.5	0.6
2	2	2.15	22.31	2.25	12.79	39.41	2.66	29.6	0.5
2	2	4.46	35.27	3.11	15.15	8.92	5.71	30.1	0.4
3	1	4.85	35.62	2.96	14.56	16.33	3.31	30.1	0.3
3	1	4.71	26.13	2.87	15.88	25.61	6.05	29.6	0.5
3	2	5.01	38	3.65	12.32	10.65	1.98	30.2	1
3	2	2.86	30.54	2.04	14.4	25.23	6.13	30.3	1.02
4	1	5.17	35.23	3.51	14.36	12.29	3.62	30	1.03
4	1	5.38	32.16	2.48	12.93	11.58	13.13	30.7	1.04
4	2	3.38	22.67	2.36	11.69	40.1	3.28	30.2	1.05
4	2	4.35	29.55	3.09	16.38	17.07	4.01	30.5	2.01
5	1	5.24	34.09	3.8	14.48	11.29	4.93	30.1	2.3
5	1	5.49	29.84	2.9	15.54	16.79	4.56	30.2	2.02
5	2	2.97	26.13	2.05	13.22	33.3	4.2	30.1	2.04
5	2	5.29	29.25	2.9	13.59	20.12	4.56	30.4	2.05

Oils: 1, control; 2, linseed; 3, soybean; 4, anchovy; and 5, tuna.
Fertilization: 1, no fertilization; 2, fertilization with urea and TSP (4 kg N and 2 kg P·day^{-1}).

Appendix A
Standard Statistical Tables

Appendix A1. Proportions of the normal curve (one-tailed).

z	0	1	2	3	4	5	6	7	8	9
0.0	0.5000	0.4960	0.4920	0.4880	0.4840	0.4801	0.4761	0.4721	0.4681	0.4641
0.1	0.4602	0.4562	0.4522	0.4483	0.4443	0.4404	0.4364	0.4325	0.4286	0.4247
0.2	0.4207	0.4168	0.4129	0.4090	0.4052	0.4013	0.3974	0.3936	0.3897	0.3859
0.3	0.3821	0.3783	0.3745	0.3707	0.3669	0.3632	0.3594	0.3557	0.3520	0.3483
0.4	0.3446	0.3409	0.3372	0.3336	0.3300	0.3264	0.3228	0.3192	0.3156	0.3121
0.5	0.3085	0.3050	0.3015	0.2981	0.2946	0.2912	0.2877	0.2843	0.2810	0.2776
0.6	0.2743	0.2709	0.2676	0.2643	0.2611	0.2578	0.2546	0.2514	0.2483	0.2451
0.7	0.2420	0.2389	0.2358	0.2327	0.2297	0.2266	0.2236	0.2207	0.2177	0.2148
0.8	0.2119	0.2090	0.2061	0.2033	0.2005	0.1977	0.1949	0.1922	0.1894	0.1867
0.9	0.1841	0.1814	0.1788	0.1762	0.1736	0.1711	0.1685	0.1660	0.1635	0.1611
1.0	0.1587	0.1562	0.1539	0.1515	0.1492	0.1469	0.1446	0.1423	0.1401	0.1379
1.1	0.1357	0.1335	0.1314	0.1292	0.1271	0.1251	0.1230	0.1210	0.1190	0.1170
1.2	0.1151	0.1131	0.1112	0.1093	0.1075	0.1056	0.1038	0.1020	0.1003	0.0985
1.3	0.0968	0.0951	0.0934	0.0918	0.0901	0.0885	0.0869	0.0853	0.0838	0.0823
1.4	0.0808	0.0793	0.0778	0.0764	0.0749	0.0735	0.0721	0.0708	0.0694	0.0681
1.5	0.0668	0.0655	0.0643	0.0630	0.0618	0.0606	0.0594	0.0582	0.0571	0.0559
1.6	0.0548	0.0537	0.0526	0.0516	0.0505	0.0495	0.0485	0.0475	0.0465	0.0455
1.7	0.0446	0.0436	0.0427	0.0418	0.0409	0.0401	0.0392	0.0384	0.0375	0.0367
1.8	0.0359	0.0351	0.0344	0.0336	0.0329	0.0322	0.0314	0.0307	0.0301	0.0294
1.9	0.0287	0.0281	0.0274	0.0268	0.0262	0.0256	0.0250	0.0244	0.0239	0.0233
2.0	0.0228	0.0222	0.0217	0.0212	0.0207	0.0202	0.0197	0.0192	0.0188	0.0183
2.1	0.0179	0.0174	0.0170	0.0166	0.0162	0.0158	0.0154	0.0150	0.0146	0.0143
2.2	0.0139	0.0136	0.0132	0.0129	0.0125	0.0122	0.0119	0.0116	0.0113	0.0110
2.3	0.0107	0.0104	0.0102	0.0099	0.0096	0.0094	0.0091	0.0089	0.0087	0.0084
2.4	0.0082	0.0080	0.0078	0.0075	0.0073	0.0071	0.0069	0.0068	0.0066	0.0064
2.5	0.0062	0.0060	0.0059	0.0057	0.0055	0.0054	0.0052	0.0051	0.0049	0.0048
2.6	0.0047	0.0045	0.0044	0.0043	0.0041	0.0040	0.0039	0.0038	0.0037	0.0036
2.7	0.0035	0.0034	0.0033	0.0032	0.0031	0.0030	0.0029	0.0028	0.0027	0.0026
2.8	0.0026	0.0025	0.0024	0.0023	0.0023	0.0022	0.0021	0.0021	0.0020	0.0019
2.9	0.0019	0.0018	0.0018	0.0017	0.0016	0.0016	0.0015	0.0015	0.0014	0.0014
3.0	0.0013	0.0013	0.0013	0.0012	0.0012	0.0011	0.0011	0.0011	0.0010	0.0010
3.1	0.0010	0.0009	0.0009	0.0009	0.0008	0.0008	0.0008	0.0008	0.0007	0.0007
3.2	0.0007	0.0007	0.0006	0.0006	0.0006	0.0006	0.0006	0.0005	0.0005	0.0005
3.3	0.0005	0.0005	0.0005	0.0004	0.0004	0.0004	0.0004	0.0004	0.0004	0.0003
3.4	0.0003	0.0003	0.0003	0.0003	0.0003	0.0003	0.0003	0.0003	0.0003	0.0002
3.5	0.0002	0.0002	0.0002	0.0002	0.0002	0.0002	0.0002	0.0002	0.0002	0.0002
3.6	0.0002	0.0002	0.0001	0.0001	0.0001	0.0001	0.0001	0.0001	0.0001	0.0001
3.7	0.0001	0.0001	0.0001	0.0001	0.0001	0.0001	0.0001	0.0001	0.0001	0.0001
3.8	0.0001	0.0001	0.0001	0.0001	0.0001	0.0001	0.0001	0.0001	0.0001	0.0001

Appendix A2. Critical values of the chi-square distribution.

Degree of freedom	Probability of Larger Value of Chi-Square (χ^2 Table).								
	0.99	0.95	0.90	0.75	0.50	0.25	0.10	0.05	0.01
1	0.000	0.004	0.016	0.102	0.455	1.323	2.706	3.841	6.635
2	0.020	0.103	0.211	0.575	1.386	2.773	4.605	5.991	9.210
3	0.115	0.352	0.584	1.213	2.366	4.108	6.251	7.815	11.345
4	0.297	0.711	1.064	1.923	3.357	5.385	7.779	9.488	13.277
5	0.554	1.145	1.610	2.675	4.351	6.626	9.236	11.070	15.086
6	0.872	1.635	2.204	3.455	5.348	7.841	10.645	12.592	16.812
7	1.239	2.167	2.833	4.255	6.346	9.037	12.017	14.067	18.475
8	1.646	2.733	3.490	5.071	7.344	10.219	13.362	15.507	20.090
9	2.088	3.325	4.168	5.899	8.343	11.389	14.684	16.919	21.666
10	2.558	3.940	4.865	6.737	9.342	12.549	15.987	18.307	23.209
11	3.053	4.575	5.578	7.584	10.341	13.701	17.275	19.675	24.725
12	3.571	5.226	6.304	8.438	11.340	14.845	18.549	21.026	26.217
13	4.107	5.892	7.042	9.299	12.340	15.984	19.812	22.362	27.688
14	4.660	6.571	7.790	10.165	13.339	17.117	21.064	23.685	29.141
15	5.229	7.261	8.547	11.037	14.339	18.245	22.307	24.996	30.578
16	5.812	7.962	9.312	11.912	15.338	19.369	23.542	26.296	32.000
17	6.408	8.672	10.085	12.792	16.338	20.489	24.769	27.587	33.409
18	7.015	9.390	10.865	13.675	17.338	21.605	25.989	28.869	34.805
19	7.633	10.117	11.651	14.562	18.338	22.718	27.204	30.144	36.191
20	8.260	10.851	12.443	15.452	19.337	23.828	28.412	31.410	37.566
22	9.542	12.338	14.041	17.240	21.337	26.039	30.813	33.924	40.289
24	10.856	13.848	15.659	19.037	23.337	28.241	33.196	36.415	42.980
26	12.198	15.379	17.292	20.843	25.336	30.435	35.563	38.885	45.642
28	13.565	16.928	18.939	22.657	27.336	32.620	37.916	41.337	48.278
30	14.953	18.493	20.599	24.478	29.336	34.800	40.256	43.773	50.892
40	22.164	26.509	29.051	33.660	39.335	45.616	51.805	55.758	63.691
50	29.707	34.764	37.689	42.942	49.335	56.334	63.167	67.505	76.154
60	37.485	43.188	46.459	52.294	59.335	66.981	74.397	79.082	88.379

Appendix A3. Critical values of d_{max} for the one-sample Kolmogorov–Smirnov (K–S) test for testing goodness of fit for discrete or grouped data (short version is provided here; for more observations please consult Zar (1984) or other statistical books.

k	n	0.20	0.10	0.05	0.02	0.01	k	n	0.20	0.10	0.05	0.02	0.01
3	3	2	2	3	3	3	4	92	7	9	10	12	13
3	6	3	3	3	4	4	4	96	7	9	10	12	13
3	9	3	4	4	4	5	4	100	8	9	11	12	13
3	12	4	4	4	5	5	5	5	3	3	3	4	4
3	15	4	4	5	6	6	5	10	3	4	4	5	5
3	18	4	5	5	6	6	5	15	4	5	5	6	6
3	21	4	5	6	6	7	5	20	5	5	6	7	7
3	24	5	5	6	7	7	5	25	5	6	6	7	8
3	27	5	6	6	7	8	5	30	5	6	7	8	8
3	30	5	6	7	8	8	5	35	6	7	7	8	9
3	33	5	6	7	8	8	5	40	6	7	8	9	10
3	35	5	6	7	8	9	5	45	6	7	8	9	10
3	39	6	7	7	8	9	5	50	7	8	9	10	11
3	42	6	7	8	9	9	5	55	7	8	9	10	11
3	45	6	7	8	9	10	5	60	7	8	9	11	12
3	48	6	7	8	9	10	5	65	7	9	10	11	12
3	51	6	7	8	10	10	5	70	8	9	10	11	12
3	54	6	8	9	10	11	5	75	8	9	10	12	13
3	57	7	8	9	10	11	5	80	8	9	11	12	13
3	60	7	8	9	10	11	5	85	8	9	11	12	13
3	63	7	8	9	10	11	5	90	8	10	11	12	13
3	66	7	8	9	10	11	5	95	8	9	11	12	13
3	69	7	8	9	11	12	5	100	8	9	11	12	14
3	72	7	8	9	11	12	6	6	3	3	4	4	4
3	75	7	8	10	11	12	6	12	4	4	5	5	6
3	78	7	9	10	11	12	6	18	4	5	6	6	7
3	81	7	9	10	11	13	6	24	5	6	6	7	8
3	84	7	9	10	12	13	6	30	6	6	7	8	9
3	87	7	9	10	12	13	6	36	6	7	8	9	9
3	90	7	9	10	12	13	6	42	6	7	8	9	10
3	93	7	9	11	12	13	6	48	7	8	9	10	11
3	96	7	9	10	12	13	6	54	7	8	9	10	11
3	99	7	9	10	12	13	6	60	7	9	10	11	12
4	4	2	3	3	3	3	6	66	8	9	10	11	12
4	8	3	4	4	4	5	6	72	8	9	10	12	13
4	12	4	4	5	5	6	6	78	8	9	11	12	13
4	16	4	5	5	6	6	6	84	8	9	11	12	13
4	20	4	5	6	6	7	6	90	8	10	11	13	14
4	24	5	6	6	7	8	6	96	8	10	11	13	14
4	28	5	6	7	7	8	7	7	3	4	4	4	5
4	32	5	6	7	8	9	7	14	4	5	5	6	6
4	36	6	7	7	8	9	7	21	5	6	6	7	7
4	40	6	7	8	9	9	7	28	5	6	7	8	8
4	44	6	7	8	9	10	7	35	6	7	8	9	9
4	48	6	7	8	10	10	7	42	6	7	8	9	10
4	52	7	8	9	10	11	7	49	7	8	9	10	11
4	56	7	8	9	10	11	7	56	7	8	9	11	12
4	60	7	8	9	10	11	7	63	8	9	10	12	12
4	64	7	8	9	11	12	7	70	8	9	10	12	13
4	68	7	9	10	11	12	7	77	8	9	11	12	13
4	72	7	9	10	11	12	7	84	8	10	12	12	13
4	76	8	9	10	11	12	7	91	8	10	11	13	14
4	80	8	9	10	11	12	7	98	8	10	11	13	14
4	84	7	9	10	12	13	8	8	3	4	4	5	5
4	88	7	9	10	12	13	8	16	4	5	6	6	7

Appendix A4. Critical values of the t-distribution (t-table).

ν	α(2): 0.50 α(1): 0.25	0.20 0.10	0.10 0.05	0.05 0.025	0.02 0.01	0.01 0.005	0.005 0.0025	0.002 0.001	0.001 0.0005
1	1.000	3.078	6.314	12.706	31.821	63.657	127.321	318.309	636.619
2	0.816	1.886	2.920	4.303	6.965	9.925	14.089	22.327	31.599
3	0.765	1.638	2.353	3.182	4.541	5.841	7.453	10.215	12.924
4	0.741	1.533	2.132	2.776	3.747	4.604	5.598	7.173	8.610
5	0.727	1.476	2.015	2.571	3.365	4.032	4.773	5.893	6.869
6	0.718	1.440	1.943	2.447	3.143	3.707	4.317	5.208	5.959
7	0.711	1.415	1.895	2.365	2.998	3.499	4.029	4.785	5.408
8	0.706	1.397	1.860	2.306	2.896	3.355	3.833	4.501	5.041
9	0.703	1.383	1.833	2.262	2.821	3.250	3.690	4.297	4.781
10	0.700	1.372	1.812	2.228	2.764	3.169	3.581	4.144	4.587
11	0.697	1.363	1.796	2.201	2.718	3.106	3.497	4.025	4.437
12	0.695	1.356	1.782	2.179	2.681	3.055	3.428	3.930	4.318
13	0.694	1.350	1.771	2.160	2.650	3.012	3.372	3.825	4.221
14	0.692	1.345	1.761	2.145	2.624	2.977	3.326	3.787	4.140
15	0.691	1.341	1.753	2.131	2.602	2.947	3.286	3.733	4.073
16	0.690	1.337	1.746	2.120	2.583	2.921	3.252	3.686	4.015
17	0.689	1.333	1.740	2.110	2.567	2.898	3.222	3.646	3.965
18	0.688	1.330	1.734	2.101	2.552	2.878	3.197	3.610	3.922
19	0.688	1.328	1.729	2.093	2.539	2.861	3.174	3.579	3.883
20	0.687	1.325	1.725	2.086	2.528	2.845	3.153	3.552	3.850
21	0.686	1.323	1.721	2.080	2.518	2.831	3.135	3.527	3.819
22	0.686	1.321	1.717	2.074	2.508	2.819	3.119	3.505	3.792
23	0.685	1.319	1.714	2.069	2.500	2.807	3.104	3.485	3.768
24	0.685	1.318	1.711	2.064	2.492	2.797	3.091	3.467	3.745
25	0.684	1.316	1.708	2.060	2.485	2.787	3.078	3.450	3.725
26	0.684	1.315	1.706	2.056	2.479	2.779	3.067	3.435	3.707
27	0.684	1.314	1.703	2.052	2.473	2.771	3.057	3.421	3.690
28	0.683	1.313	1.701	2.048	2.467	2.763	3.047	3.408	3.674
29	0.683	1.311	1.699	2.045	2.462	2.756	3.038	3.396	3.659
30	0.683	1.310	1.697	2.042	2.457	2.750	3.030	3.385	3.646
31	0.682	1.309	1.696	2.040	2.453	2.744	3.022	3.375	3.633
32	0.682	1.309	1.694	2.037	2.449	2.738	3.015	3.365	3.622
33	0.682	1.308	1.692	2.035	2.445	2.733	3.008	3.356	3.611
34	0.682	1.307	1.691	2.032	2.441	2.728	3.002	3.348	3.601
35	0.682	1.306	1.690	2.030	2.438	2.724	2.996	3.340	3.591
36	0.681	1.306	1.688	2.028	2.434	2.719	2.990	3.333	3.582
37	0.681	1.305	1.687	2.026	2.431	2.715	2.985	3.326	3.574
38	0.681	1.304	1.686	2.024	2.429	2.712	2.980	3.319	3.566
39	0.681	1.304	1.685	2.023	2.426	2.708	2.976	3.313	3.558
40	0.681	1.303	1.684	2.021	2.423	2.704	2.971	3.307	3.551
50	0.679	1.299	1.671	2.009	2.403	2.678	2.937	3.261	3.496
60	0.679	1.296	1.676	2.000	2.390	2.660	2.915	3.232	3.460
70	0.678	1.294	1.667	1.994	2.381	2.648	2.899	3.211	3.435
80	0.678	1.292	1.664	1.990	2.374	2.639	2.887	3.195	3.416
90	0.677	1.291	1.662	1.987	2.368	2.632	2.878	3.183	3.402
100	0.677	1.290	1.660	1.984	2.364	2.626	2.871	3.174	3.390
500	0.675	1.283	1.648	1.965	2.334	2.586	2.820	3.107	3.310
600	0.675	1.283	1.647	1.964	2.333	2.584	2.817	3.104	3.307
1,000	0.675	1.283	1.646	1.962	2.330	2.581	2.813	3.098	3.300
∞	0.6745	1.2816	1.6449	1.9600	2.3263	2.5758	2.8070	3.0902	3.2905

Appendix A5. U-table for Mann–Whitney test.

n_1 Degree of freedom (for greater mean square)

n_2	1	2	3	4	5	6	7	8	9	10	11	12	13	14	15	16	17	18	19	20
3	—	—	—	—	—	—	—	—	—	—	—	—	—	—	—	—	—	—	—	—
4	—	—	—	16	—	—	—	—	—	—	—	—	—	—	—	—	—	—	—	—
5	—	—	15	19	23	—	—	—	—	—	—	—	—	—	—	—	—	—	—	—
	—	—	—	—	25	—	—	—	—	—	—	—	—	—	—	—	—	—	—	—
6	—	—	17	22	27	31	—	—	—	—	—	—	—	—	—	—	—	—	—	—
	—	—	—	24	29	34	—	—	—	—	—	—	—	—	—	—	—	—	—	—
7	—	—	20	25	30	36	41	—	—	—	—	—	—	—	—	—	—	—	—	—
	—	—	—	28	34	39	45	—	—	—	—	—	—	—	—	—	—	—	—	—
8	—	16	22	28	34	40	46	51	—	—	—	—	—	—	—	—	—	—	—	—
	—	—	—	31	38	44	50	57	—	—	—	—	—	—	—	—	—	—	—	—
9	—	18	25	32	38	44	51	57	64	—	—	—	—	—	—	—	—	—	—	—
	—	—	27	35	42	49	56	63	70	—	—	—	—	—	—	—	—	—	—	—
10	—	20	27	35	42	49	56	63	70	77	—	—	—	—	—	—	—	—	—	—
	—	—	30	38	46	54	61	69	77	84	—	—	—	—	—	—	—	—	—	—
11	—	22	30	38	46	53	61	69	76	84	91	—	—	—	—	—	—	—	—	—
	—	—	33	42	50	59	67	75	83	92	100	—	—	—	—	—	—	—	—	—
12	—	23	32	41	49	58	66	74	82	91	99	107	—	—	—	—	—	—	—	—
	—	—	35	45	54	63	72	81	90	99	108	117	—	—	—	—	—	—	—	—
15	—	29	40	50	61	71	81	91	101	111	121	131	141	151	161	—	—	—	—	—
	—	—	43	55	67	78	89	100	111	121	132	143	153	164	174	—	—	—	—	—
20	—	38	52	66	80	93	106	119	132	145	158	171	184	197	210	222	235	248	261	273
	—	40	57	72	87	102	116	130	144	158	172	186	200	213	227	241	251	268	281	295
25	—	47	65	82	98	115	131	147	163	179	195	211	227	243	258	274	290	305	321	337
	—	50	70	90	108	126	143	161	178	195	212	229	246	263	279	296	313	329	346	362
30	—	55	77	97	117	137	156	175	194	213	232	251	270	289	307	326	344	363	381	400
	—	59	84	107	128	150	170	191	212	232	252	272	292	312	331	351	371	390	410	430
35	—	64	89	113	136	159	181	203	226	247	269	291	313	334	356	377	399	420	441	463
	—	69	97	124	149	173	198	221	245	268	292	315	338	361	383	406	429	451	474	497
40	—	73	102	129	155	181	206	231	257	281	306	331	355	380	404	429	453	477	502	526
	—	78	111	141	169	197	225	252	279	305	331	358	384	410	435	461	487	512	538	563

Appendix A6. Critical values of the Wilcoxon t-distribution.

n	α(2): 0.50 α(1): 0.25	0.20 0.10	0.10 0.05	0.05 0.025	0.02 0.01	0.01 0.005	0.005 0.0025	0.001 0.0005
4	2	0						
5	4	2	0					
6	6	3	2	0				
7	9	5	3	2	0			
8	12	8	5	3	1	0		
9	16	10	8	5	3	1	0	
10	20	14	10	8	5	3	1	
11	24	17	13	10	7	5	3	0
12	29	21	17	13	9	7	5	1
13	35	26	21	17	12	9	7	2
14	40	31	25	21	15	12	9	4
15	47	36	30	25	19	15	12	6
16	54	42	35	29	23	19	15	8
17	61	48	41	34	27	23	19	11
18	69	55	47	40	32	27	23	14
19	77	62	53	46	37	32	27	18
20	86	69	60	52	43	37	32	21
21	95	77	67	58	49	42	37	25
22	104	86	75	65	55	48	42	30
23	114	94	83	73	62	54	48	35
24	125	104	91	81	69	61	54	40
25	136	113	100	89	76	68	60	45
26	148	124	110	98	84	75	67	51
27	160	134	119	107	92	83	74	57
28	172	145	130	116	101	91	82	64
29	185	157	140	126	110	100	90	71
30	198	169	151	137	120	109	98	78
31	212	181	163	147	130	118	107	86
32	226	194	175	159	140	128	116	94
33	241	207	187	170	151	138	126	102
34	257	221	200	182	162	148	136	111
35	272	235	213	195	173	159	146	120
36	289	250	227	208	185	171	157	130
37	305	265	241	221	198	182	168	140
38	323	281	256	235	211	194	180	150
39	340	297	271	249	224	207	192	161
40	358	313	286	264	238	220	204	172
41	377	330	302	279	252	233	217	183
42	396	348	319	294	266	247	230	195
43	416	365	336	310	281	261	244	207
44	436	384	353	327	296	276	258	220
45	456	402	371	343	312	291	272	233
46	477	422	389	361	328	307	287	246
47	499	441	407	378	345	322	302	260
48	521	462	426	396	362	339	318	274
49	543	482	446	415	379	355	334	289
50	566	503	466	434	397	373	350	304
51	590	525	486	453	416	390	367	319
52	613	547	507	473	434	408	384	335
53	638	569	529	494	454	427	402	351
54	668	592	550	514	473	445	420	368
55	688	615	573	536	493	465	438	385
56	714	639	595	557	514	484	457	402
57	740	664	618	579	535	504	477	420
58	767	688	642	602	556	525	497	438
59	794	714	666	625	578	546	517	457
60	822	739	690	648	600	567	537	476

Appendix A7. Points for the F-distribution (5%, light type; 1%, bold face type).

f_2	19 f_1 Degree of freedom (for greater mean square)

f_2	1	2	3	4	5	6	7	8	9	10	12	15	20	24	30	40	60	120	∞
1	161.4	199.5	215.7	224.6	230.2	234.0	236.8	238.9	240.5	241.9	243.9	245.9	248.0	249.1	250.1	251.1	252.2	253.3	254.3
	4.052	**4.9995**	**5.403**	**5.625**	**5.764**	**5.859**	**5.928**	**5.982**	**6.022**	**6.056**	**6.106**	**6.157**	**6.209**	**6.235**	**6.261**	**6.287**	**6.313**	**6.339**	**6.366**
2	18.51	19.00	19.16	19.25	19.30	19.33	19.35	19.37	19.38	19.40	19.41	19.43	19.45	19.45	19.46	19.47	19.48	19.49	19.50
	98.50	**99.00**	**99.17**	**99.25**	**99.30**	**99.33**	**99.36**	**99.37**	**99.39**	**99.40**	**99.42**	**99.43**	**99.45**	**99.46**	**99.47**	**99.47**	**99.48**	**99.49**	**99.50**
3	10.13	9.55	9.28	9.12	9.01	8.94	8.89	8.85	8.81	8.79	8.74	8.70	8.66	8.64	8.62	8.59	8.57	8.55	8.53
	34.12	**30.82**	**29.46**	**28.71**	**28.24**	**27.91**	**27.67**	**27.49**	**27.3**	**27.23**	**27.05**	**26.87**	**26.69**	**26.60**	**26.50**	**26.41**	**26.32**	**26.22**	**26.13**
4	7.71	6.94	6.59	6.39	6.26	6.16	6.09	6.04	6.00	5.96	5.91	5.86	5.80	5.77	5.75	5.72	5.69	5.66	5.63
	21.20	**18.00**	**16.69**	**15.98**	**15.52**	**15.21**	**14.98**	**14.80**	**14.66**	**14.55**	**14.37**	**14.20**	**14.02**	**13.93**	**13.84**	**13.75**	**13.65**	**13.56**	**13.46**
5	6.61	5.79	5.41	5.19	5.05	4.95	4.88	4.82	4.77	4.74	4.68	4.62	4.56	4.53	4.50	4.46	4.43	4.40	4.36
	16.26	**13.27**	**12.06**	**11.39**	**10.97**	**10.67**	**10.46**	**10.29**	**10.16**	**10.05**	**9.89**	**9.72**	**9.55**	**9.47**	**9.38**	**9.29**	**9.20**	**9.11**	**9.02**
6	5.99	5.14	4.76	4.53	4.39	4.28	4.21	4.15	4.10	4.06	4.00	3.94	3.87	3.84	3.81	3.77	3.74	3.70	3.67
	13.74	**10.92**	**9.78**	**9.15**	**8.75**	**8.47**	**8.26**	**8.10**	**7.98**	**7.87**	**7.72**	**7.56**	**7.40**	**7.31**	**7.23**	**7.14**	**7.06**	**6.97**	**6.88**
7	5.59	4.74	4.35	4.12	3.97	3.87	3.79	3.73	3.68	3.64	3.57	3.51	3.44	3.41	3.38	3.34	3.30	3.27	3.23
	12.25	**9.55**	**8.45**	**7.85**	**7.46**	**7.19**	**6.99**	**6.84**	**6.72**	**6.62**	**6.47**	**6.31**	**6.16**	**6.07**	**5.99**	**5.91**	**5.82**	**5.74**	**5.65**
8	5.32	4.46	4.07	3.84	3.69	3.58	3.50	3.44	3.39	3.35	3.28	3.22	3.15	3.12	3.08	3.05	3.01	2.97	2.93
	11.26	**8.65**	**7.59**	**7.01**	**6.63**	**6.37**	**6.18**	**6.03**	**5.91**	**5.81**	**5.67**	**5.52**	**5.36**	**5.28**	**5.20**	**5.12**	**5.03**	**4.95**	**4.86**
9	5.12	4.26	3.86	3.63	3.48	3.37	3.29	3.23	3.18	3.14	3.07	3.01	2.94	2.90	2.86	2.83	2.79	2.75	2.71
	10.56	**8.02**	**6.99**	**6.42**	**6.06**	**5.80**	**5.61**	**5.47**	**5.35**	**5.26**	**5.11**	**4.96**	**4.81**	**4.73**	**4.65**	**4.57**	**4.48**	**4.40**	**4.31**
10	4.96	4.10	3.71	3.48	3.33	3.22	3.14	3.07	3.02	2.98	2.91	2.85	2.77	2.74	2.70	2.66	2.62	2.58	2.54
	10.04	**7.56**	**6.55**	**5.99**	**5.64**	**5.39**	**5.20**	**5.06**	**4.94**	**4.85**	**4.71**	**4.56**	**4.41**	**4.33**	**4.25**	**4.17**	**4.08**	**4.00**	**3.91**
11	4.84	3.98	3.59	3.36	3.20	3.09	3.01	2.95	2.90	2.85	2.79	2.72	2.65	2.61	2.57	2.53	2.49	2.45	2.40
	9.65	**7.21**	**6.22**	**5.67**	**5.32**	**5.07**	**4.89**	**4.74**	**4.63**	**4.54**	**4.40**	**4.25**	**4.10**	**4.02**	**3.94**	**3.86**	**3.78**	**3.69**	**3.60**
12	4.75	3.89	3.49	3.26	3.11	3.00	2.91	2.85	2.80	2.75	2.69	2.62	2.54	2.51	2.47	2.43	2.38	2.34	2.30
	9.33	**6.93**	**5.95**	**5.41**	**5.06**	**4.82**	**4.64**	**4.50**	**4.39**	**4.30**	**4.16**	**4.01**	**3.86**	**3.78**	**3.70**	**3.62**	**3.54**	**3.45**	**3.36**
13	4.67	3.81	3.41	3.18	3.03	2.92	2.83	2.77	2.71	2.67	2.60	2.53	2.46	2.42	2.38	2.34	2.30	2.25	2.21
	9.07	**6.70**	**5.74**	**5.21**	**4.86**	**4.62**	**4.44**	**4.30**	**4.19**	**4.10**	**3.96**	**3.82**	**3.66**	**3.59**	**3.51**	**3.43**	**3.34**	**3.25**	**3.17**
14	4.60	3.74	3.34	3.11	2.96	2.85	2.76	2.70	2.65	2.60	2.53	2.46	2.39	2.35	2.31	2.27	2.22	2.18	2.13
	8.86	**6.51**	**5.56**	**5.04**	**4.69**	**4.46**	**4.28**	**4.14**	**4.03**	**3.94**	**3.80**	**3.66**	**3.51**	**3.43**	**3.35**	**3.27**	**3.18**	**3.09**	**3.00**

15	4.54	3.68	3.29	3.06	2.90	2.79	2.71	2.64	2.59	2.54	2.48	2.40	2.33	2.29	2.25	2.20	2.16	2.11	2.07
	8.68	**6.36**	**5.42**	**4.89**	**4.56**	**4.32**	**4.14**	**4.00**	**3.89**	**3.80**	**3.67**	**3.52**	**3.37**	**3.29**	**3.21**	**3.13**	**3.05**	**2.96**	**2.87**
16	4.49	3.63	3.24	3.01	2.85	2.74	2.66	2.59	2.54	2.49	2.42	2.35	2.28	2.24	2.19	2.15	2.11	2.06	2.01
	8.53	**6.23**	**5.29**	**4.77**	**4.44**	**4.20**	**4.03**	**3.89**	**3.78**	**3.69**	**3.55**	**3.41**	**3.26**	**3.18**	**3.10**	**3.02**	**2.93**	**2.84**	**2.75**
17	4.45	3.59	3.20	2.96	2.81	2.70	2.61	2.55	2.49	2.45	2.38	2.31	2.23	2.19	2.15	2.10	2.06	2.01	1.96
	8.40	**6.11**	**5.18**	**4.67**	**4.34**	**4.10**	**3.93**	**3.79**	**3.68**	**3.59**	**3.46**	**3.31**	**3.16**	**3.08**	**3.00**	**2.92**	**2.83**	**2.75**	**2.65**
18	4.41	3.55	3.16	2.93	2.77	2.66	2.58	2.51	2.46	2.41	2.34	2.27	2.19	2.15	2.11	2.06	2.02	1.97	1.92
	8.29	**6.01**	**5.09**	**4.58**	**4.25**	**4.01**	**3.84**	**3.71**	**3.60**	**3.51**	**3.37**	**3.23**	**3.08**	**3.00**	**2.92**	**2.84**	**2.75**	**2.66**	**2.57**
19	4.38	3.52	3.13	2.90	2.74	2.63	2.54	2.48	2.42	2.38	2.31	2.23	2.16	2.11	2.07	2.03	1.98	1.93	1.88
	8.18	**5.93**	**5.01**	**4.50**	**4.17**	**3.94**	**3.77**	**3.63**	**3.52**	**3.43**	**3.30**	**3.15**	**3.00**	**2.92**	**2.84**	**2.76**	**2.67**	**2.58**	**2.49**
20	4.35	3.49	3.10	2.87	2.71	2.60	2.51	2.45	2.39	2.35	2.28	2.20	2.12	2.08	2.04	1.99	1.95	1.90	1.84
	8.10	**5.85**	**4.94**	**4.43**	**4.10**	**3.87**	**3.70**	**3.56**	**3.46**	**3.37**	**3.23**	**3.09**	**2.94**	**2.86**	**2.78**	**2.69**	**2.61**	**2.52**	**2.42**
21	4.32	3.47	3.07	2.84	2.68	2.57	2.49	2.42	2.37	2.32	2.25	2.18	2.10	2.05	2.01	1.96	1.92	1.87	1.81
	8.02	**5.78**	**4.87**	**4.37**	**4.04**	**3.81**	**3.64**	**3.51**	**3.40**	**3.31**	**3.17**	**3.03**	**2.88**	**2.80**	**2.72**	**2.64**	**2.55**	**2.46**	**2.36**
22	4.30	3.44	3.05	2.82	2.66	2.55	2.46	2.40	2.34	2.30	2.23	2.15	2.07	2.03	1.98	1.94	1.89	1.84	1.78
	7.95	**5.72**	**4.82**	**4.31**	**3.99**	**3.76**	**3.59**	**3.45**	**3.35**	**3.26**	**3.12**	**2.98**	**2.83**	**2.75**	**2.67**	**2.58**	**2.50**	**2.40**	**2.31**
23	4.28	3.42	3.03	2.80	2.64	2.53	2.44	2.37	2.32	2.27	2.20	2.13	2.05	2.01	1.96	1.91	1.86	1.81	1.76
	7.88	**5.66**	**4.76**	**4.26**	**3.94**	**3.71**	**3.54**	**3.41**	**3.30**	**3.21**	**3.07**	**2.93**	**2.78**	**2.70**	**2.62**	**2.54**	**2.45**	**2.35**	**2.26**
24	4.26	3.40	3.01	2.78	2.62	2.51	2.42	2.36	2.30	2.25	2.18	2.11	2.03	1.98	1.94	1.89	1.84	1.79	1.73
	7.82	**5.61**	**4.72**	**4.22**	**3.90**	**3.67**	**3.50**	**3.36**	**3.26**	**3.17**	**3.03**	**2.89**	**2.74**	**2.66**	**2.58**	**2.49**	**2.40**	**2.31**	**2.21**
25	4.24	3.39	2.99	2.76	2.60	2.49	2.40	2.34	2.28	2.24	2.16	2.09	2.01	1.96	1.92	1.87	1.82	1.77	1.71
	7.77	**5.57**	**4.68**	**4.18**	**3.85**	**3.63**	**3.46**	**3.32**	**3.22**	**3.13**	**2.99**	**2.85**	**2.70**	**2.62**	**2.54**	**2.45**	**2.36**	**2.27**	**2.17**
30	4.17	3.32	2.92	2.69	2.53	2.42	2.33	2.27	2.21	2.16	2.09	2.01	1.93	1.89	1.84	1.79	1.74	1.68	1.62
	7.56	**5.39**	**4.51**	**4.02**	**3.70**	**3.47**	**3.30**	**3.17**	**3.07**	**2.98**	**2.84**	**2.70**	**2.55**	**2.47**	**2.39**	**2.30**	**2.21**	**2.11**	**2.01**
40	4.08	3.23	2.84	2.61	2.45	2.34	2.25	2.18	2.12	2.08	2.00	1.92	1.84	1.79	1.74	1.69	1.64	1.58	1.51
	7.31	**5.18**	**4.31**	**3.83**	**3.51**	**3.29**	**3.12**	**2.99**	**2.89**	**2.80**	**2.66**	**2.52**	**2.37**	**2.29**	**2.20**	**2.11**	**2.02**	**1.92**	**1.80**
60	4.00	3.15	2.76	2.53	2.37	2.25	2.17	2.10	2.04	1.99	1.92	1.84	1.75	1.70	1.65	1.59	1.53	1.47	1.39
	7.08	**4.98**	**4.13**	**3.65**	**3.34**	**3.12**	**2.95**	**2.82**	**2.72**	**2.63**	**2.50**	**2.35**	**2.20**	**2.12**	**2.03**	**1.94**	**1.84**	**1.73**	**1.60**
120	3.92	3.07	2.68	2.45	2.29	2.17	2.09	2.02	1.96	1.91	1.83	1.75	1.66	1.61	1.55	1.50	1.43	1.35	1.25
	6.85	**4.79**	**3.95**	**3.48**	**3.17**	**2.96**	**2.79**	**2.66**	**2.56**	**2.47**	**2.34**	**2.19**	**2.03**	**1.95**	**1.86**	**1.76**	**1.66**	**1.53**	**1.38**
∞	3.84	3.00	2.60	2.37	2.21	2.10	2.01	1.94	1.88	1.83	1.75	1.67	1.57	1.52	1.46	1.39	1.32	1.22	1.00
	6.63	**4.61**	**3.78**	**3.32**	**3.02**	**2.80**	**2.64**	**2.51**	**2.41**	**2.32**	**2.18**	**2.04**	**1.88**	**1.79**	**1.70**	**1.59**	**1.47**	**1.32**	**1.00**

Appendix A8. Critical values of the Kruskal–Wallis H-distribution.

n_1	n_2	n_3			$\alpha : 0.01$	0.05	0.02	0.01	0.005	0.002	0.001
3	2	2	2		5.544	6.333	6.978	7.133	7.533		
3	3	1	1		5.333	6.333					
3	3	2	1		5.689	6.244	6.689	7.200	7.400		
3	3	2	2		5.745	6.527	7.182	7.636	7.873	8.018	8.455
3	3	3	1		5.655	6.600	7.109	7.400	8.055	8.345	
3	3	3	2		5.879	6.727	7.636	8.105	8.379	8.803	9.030
3	3	3	3		6.026	7.000	7.872	8.538	8.897	9.462	9.513
4	1	1	1		–						
4	2	1	1		5.250	5.833					
4	2	2	1		5.533	6.133	6.667	7.000			
4	2	2	2		5.755	6.545	7.091	7.391	7.964	8.291	
4	3	1	1		5.067	6.178	6.711	7.067			
4	3	2	1		5.591	6.309	7.018	7.455	7.773	8.182	
4	3	2	2		5.750	6.621	7.530	7.871	8.273	8.689	8.909
4	3	3	1		5.589	6.545	7.485	7.758	8.212	8.697	9.182
4	3	3	2		5.872	6.795	7.763	8.333	8.718	9.167	9.455
4	3	3	3		6.016	6.984	7.995	8.659	9.253	9.709	10.016
4	4	1	1		5.182	5.945	7.091	7.909	7.909		
4	4	2	1		5.568	6.386	7.364	7.886	8.341	8.591	8.909
4	4	2	2		5.808	6.731	7.750	8.346	8.692	9.269	9.462
4	4	3	1		5.692	6.635	7.660	8.231	8.583	9.038	9.327
4	4	3	2		5.901	6.874	7.951	8.621	9.165	9.615	9.945
4	4	3	3		6.019	7.038	8.181	8.876	9.495	10.105	10.467
4	4	4	1		5.564	6.725	7.879	8.588	9.000	9.478	9.758
4	4	4	2		5.914	6.957	8.157	8.871	9.486	10.043	10.429
4	4	4	3		6.042	7.142	8.350	9.075	9.742	10.542	10.929
4	4	4	4		6.088	7.235	8.515	9.287	9.971	10.809	11.338
2	1	1	1	1	–						
2	2	1	1	1	5.786						
2	2	2	1	1	6.250	6.750					
2	2	2	2	1	6.600	7.133	(7.533)	7.533			
2	2	2	2	2	6.982	7.418	8.073	8.291	(8.727)	8.727	
3	1	1	1	1	–						
3	2	1	1	1	6.139	6.583					
3	2	2	1	1	6.511	6.800	7.400	7.600			
3	2	2	2	1	6.709	7.309	7.836	8.127	8.327	8.618	
3	2	2	2	2	6.955	7.682	8.303	8.682	8.985	9.273	9.364
3	3	1	1	1	6.311	7.111	7.467				
3	3	2	1	1	6.600	7.200	7.892	8.073	8.345		
3	3	2	2	1	6.788	7.591	8.258	8.576	8.924	9.167	9.303
3	3	2	2	2	7.026	7.910	8.667	9.115	9.474	9.769	10.026
3	3	3	1	1	6.788	7.576	8.242	8.424	8.848	(9.455)	9.455
3	3	3	2	1	6.910	7.769	8.590	9.051	9.410	9.769	9.974
3	3	3	2	2	7.121	8.044	9.011	9.505	9.890	10.330	10.637
3	3	3	3	1	7.077	8.000	8.879	9.451	9.846	10.286	10.549
3	3	3	3	2	7.210	8.200	9.267	9.876	10.333	10.838	11.171
3	3	3	3	3	7.333	8.333	9.467	10.200	10.733	10.267	11.667

Appendix A9. Critical values of the Friedman χ_r^2 distribution.

a (*n*)	*b* (*m*)	α : 0.50	0.20	0.10	0.05	0.02	0.01	0.005	0.002	0.001
3	2	3.000	4.000							
3	3	2.667	4.667	(6.000)	6.000					
3	4	2.000	4.500	6.000	6.500	(8.000)	(8.000)	8.000		
3	5	2.800	3.600	5.200	6.400	(8.400)	8.400	(10.000)	(10.000)	10.000
3	6	2.330	4.000	5.330	7.000	8.330	9.000	(10.330)	10.330	12.000
3	7	2.000	3.714	5.429	7.143	8.000	8.857	10.286	11.143	12.286
3	8	2.250	4.000	5.250	6.250	7.750	9.000	9.750	12.000	12.250
3	9	2.000	3.556	5.556	6.222	8.000	9.556	10.667	11.556	12.667
3	10	1.800	3.800	5.000	6.200	7.800	9.600	10.400	12.200	12.600
3	11	4.636	3.818	4.909	6.545	7.818	9.455	10.364	11.636	13.273
3	12	1.500	3.500	5.167	6.167	8.000	9.500	10.167	12.167	12.500
3	13	1.846	3.846	4.769	6.000	8.000	9.385	10.308	11.538	12.923
3	14	1.714	3.571	5.143	6.143	8.143	9.000	10.429	12.000	13.286
3	15	1.733	3.600	4.933	6.400	8.133	8.933	10.000	12.133	12.933
4	2	3.600	5.400	(6.000)	6.000					
4	3	3.400	5.400	6.600	7.400	8.200	(9.000)	(9.000)	9.000	
4	4	3.000	4.800	6.300	7.800	8.400	9.600	(10.200)	10.200	11.100
4	5	3.000	5.160	6.360	7.800	9.240	9.960	10.920	11.640	12.600
4	6	3.000	4.800	6.400	7.600	9.400	10.200	11.400	12.200	12.800
4	7	2.829	4.886	6.429	7.800	9.343	10..371	11.400	12.771	13.800
4	8	2.550	4.800	6.300	7.650	9.450	10.350	11.850	12.900	13.800

Appendix A10. Critical values of the Spearman's rank correlation coefficient, r_s.

n	α(2): 0.50 / α(1): 0.25	0.20 / 0.10	0.10 / 0.05	0.05 / 0.025	0.02 / 0.01	0.01 / 0.005	0.005 / 0.0025	0.002 / 0.001	0.001 / 0.0005
4	0.600	1.000	1.000						
5	0.500	0.800	0.900	1.000	1.000				
6	0.371	0.657	0.829	0.886	0.943	1.000	1.000		
7	0.321	0.571	0.714	0.786	0.893	0.929	0.964	1.000	1.000
8	0.310	0.524	0.643	0.738	0.833	0.881	0.905	0.952	0.976
9	0.267	0.483	0.600	0.700	0.783	0.833	0.867	0.917	0.933
10	0.248	0.455	0.564	0.648	0.745	0.794	0.830	0.879	0.903
11	0.236	0.427	0.536	0.618	0.709	0.755	0.800	0.845	0.873
12	0.217	0.406	0.503	0.587	0.678	0.727	0.769	0.818	0.846
13	0.209	0.385	0.484	0.560	0.648	0.703	0.747	0.791	0.824
14	0.200	0.367	0.464	0.538	0.626	0.679	0.723	0.771	0.802
15	0.189	0.354	0.446	0.521	0.604	0.654	0.700	0.750	0.779
16	0.182	0.341	0.429	0.503	0.582	0.635	0.679	0.729	0.762
17	0.176	0.328	0.414	0.485	0.566	0.615	0.662	0.713	0.748
18	0.170	0.317	0.401	0.472	0.550	0.600	0.643	0.695	0.728
19	0.165	0.309	0.391	0.460	0.535	0.584	0.628	0.677	0.712
20	0.161	0.299	0.380	0.447	0.520	0.570	0.612	0.662	0.696
21	0.156	0.292	0.370	0.435	0.508	0.556	0.599	0.648	0.681
22	0.152	0.284	0.361	0.425	0.496	0.544	0.586	0.634	0.667
23	0.148	0.278	0.353	0.415	0.486	0.532	0.573	0.622	0.654
24	0.144	0.271	0.344	0.406	0.476	0.521	0.562	0.610	0.642
25	0.142	0.265	0.337	0.398	0.466	0.511	0.551	0.598	0.630
26	0.138	0.259	0.331	0.390	0.457	0.501	0.541	0.587	0.619
27	0.136	0.255	0.324	0.382	0.448	0.491	0.531	0.577	0.608
28	0.133	0.250	0.317	0.375	0.440	0.483	0.522	0.567	0.598
29	0.130	0.245	0.312	0.368	0.433	0.475	0.513	0.558	0.589
30	0.128	0.240	0.306	0.362	0.425	0.467	0.504	0.549	0.580
31	0.126	0.236	0.301	0.356	0.418	0.459	0.496	0.541	0.571
32	0.124	0.232	0.296	0.350	0.412	0.452	0.489	0.533	0.563
33	0.121	0.229	0.291	0.345	0.405	0.446	0.482	0.525	0.554
34	0.120	0.225	0.287	0.340	0.399	0.439	0.475	0.517	0.547
35	0.118	0.222	0.283	0.335	0.394	0.433	0.468	0.510	0.539
36	0.116	0.219	0.279	0.330	0.388	0.427	0.462	0.504	0.533
37	0.114	0.216	0.275	0.325	0.383	0.421	0.456	0.497	0.526
38	0.113	0.212	0.271	0.321	0.378	0.415	0.450	0.491	0.519
39	0.111	0.210	0.267	0.317	0.373	0.410	0.444	0.485	0.513
40	0.110	0.207	0.264	0.313	0.368	0.405	0.439	0.479	0.507
41	0.108	0.204	0.261	0.309	0.364	0.400	0.433	0.473	0.501
42	0.107	0.202	0.257	0.305	0.359	0.395	0.428	0.468	0.495
43	0.105	0.199	0.254	0.301	0.355	0.391	0.423	0.463	0.490
44	0.104	0.197	0.251	0.298	0.351	0.386	0.419	0.458	0.484
45	0.103	0.194	0.248	0.294	0.347	0.382	0.414	0.453	0.479
46	0.102	0.192	0.246	0.291	0.343	0.378	0.410	0.448	0.474
47	0.101	0.190	0.243	0.288	0.340	0.374	0.405	0.443	0.469
48	0.100	0.188	0.240	0.285	0.336	0.370	0.401	0.439	0.465
49	0.098	0.186	0.238	0.282	0.333	0.366	0.397	0.434	0.460
50	0.097	0.184	0.235	0.279	0.329	0.363	0.393	0.430	0.456
51	0.096	0.182	0.233	0.276	0.326	0.359	0.390	0.426	0.451
52	0.095	0.180	0.231	0.274	0.323	0.356	0.386	0.422	0.447
53	0.095	0.179	0.228	0.271	0.320	0.352	0.382	0.418	0.443
54	0.094	0.177	0.226	0.268	0.317	0.349	0.379	0.414	0.439
55	0.093	0.175	0.224	0.266	0.314	0.346	0.375	0.411	0.435

Data arrangement and analysis

Appendix B1. Three factors with 14 response variables arranged for statistical analysis.

Hapa No.	Feeding Rate	Hap-Exch	S-Day	FTWT	FNO	FSURV	MTWT	MNO	MSURV	FMWT	MMWT	DENSITY	BIOMASS	SEEDNO1	SEEDNO2	SEEDWT	TEMP3	DO6
1	1.5	5	0	35.7	360.0	100	47.0	360.0	100	99.2	130.6	6.0	0.69				34.3	3.4
1	1.5	5	10	32.0	320.0	89	42.4	320.0	89	100.0	132.5	5.3	0.62	86,431	0.86	382	27.2	4.4
1	1.5	5	25	36.4	318.0	88	47.6	293.0	81	114.5	162.5	5.1	0.70	113,593	1.14	511	32.2	1.5
1	1.5	5	30	37.2	325.0	90	46.4	292.0	81	114.5	158.9	5.1	0.70	51,813	0.52	240	30.1	3.6
1	1.5	5	50	39.0	285.0	79	49.2	254.0	71	136.8	193.7	4.5	0.74	212,876	2.13	937	32	2.1
1	1.5	5	65	40.0	272.0	76	52.8	275.0	76	147.1	192.0	4.6	0.77	198,324	1.98	910	31.4	2.7
1	1.5	5	80	46.8	245.0	68	56.1	263.0	73	191.0	213.3	4.2	0.86	181,355	1.81	850	31.7	1.7
1	1.5	5	95	51.6	262.0	73	60.3	254.0	71	196.9	237.4	4.3	0.93	162,478	1.62	785	33.2	0.9
1	1.5	5	110	56.4	275.0	76	65.2	255.0	71	205.1	255.7	4.4	1.01	112,361	1.12	549	33.45	1.05
1	1.5	5	120	45.0	256.0	71	50.8	255.0	71	175.8	199.2	4.3	0.80	113,327	1.13	505	33.7	1.2
2	0.75	60	0	38.7	360.0	100	48.6	360.0	100	107.5	135.0	6.0	0.73				34.3	3.1
2	0.75	60	10	33.0	320.0	89	42.2	320.0	89	103.1	131.9	5.3	0.63	54,128	0.54	250	27	4.8
2	0.75	60	25	38.0	362.0	101	42.7	286.0	79	105.0	149.3	5.4	0.67	138,837	1.39	629	32.2	1.1
2	0.75	60	30	36.4	319.0	89	41.8	296.0	82	114.1	141.2	5.1	0.65	51,245	0.51	230	29.8	3.5
2	0.75	60	50	38.4	334.0	93	40.8	271.0	75	115.0	150.6	5.0	0.66	307,014	3.07	1,363	31.3	2.8
2	0.75	60	65	37.8	324.0	90	41.6	246.0	68	116.7	169.1	4.8	0.66	228,620	2.29	1,062	31.1	2.7
2	0.75	60	80	41.8	331.0	92	42.9	244.0	68	126.3	175.8	4.8	0.71	221,956	2.22	1,031	30.8	1.7
2	0.75	60	95	44.0	312.0	87	46.4	251.0	70	141.0	184.9	4.7	0.75	172,089	1.72	830	32.4	1
2	0.75	60	110	44.2	304.0	84	47.6	248.0	69	145.4	191.9	4.6	0.77	188,756	1.89	910	32.75	1.2
2	0.75	60	120	56.0	311.0	86	45.0	252.0	70	180.1	178.6	4.7	0.84	196,037	1.96	832	33.1	1.4
3	1.5	15	0	38.2	360.0	100	50.5	360.0	100	106.1	140.3	6.0	0.74				33.8	3.1
3	1.5	15	10	30.5	320.0	89	31.0	320.0	89	95.3	96.9	5.3	0.51	37,554	0.38	171	27.2	4.3
3	1.5	15	25	39.2	376.0	104	44.0	268.0	74	104.3	164.2	5.4	0.69	166,179	1.66	748	32.1	1.2
3	1.5	15	30	34.9	320.0	89	44.7	294.0	82	109.1	152.0	5.1	0.66	54,437	0.54	246	29.8	2.7
3	1.5	15	50	37.0	304.0	84	48.2	273.0	76	121.7	176.6	4.8	0.71	314,733	3.15	1,391	31.2	2.4
3	1.5	15	65	40.8	282.0	78	52.8	238.0	66	144.7	221.8	4.3	0.78	250,493	2.50	1,157	30.9	2.8
3	1.5	15	80	42.2	255.0	71	55.8	241.0	67	165.5	231.5	4.1	0.82	206,382	2.06	989	31.1	1.7
3	1.5	15	95	48.4	270.0	75	62.0	230.0	64	179.3	269.6	4.2	0.92	194,885	1.95	927	33	0.8
3	1.5	15	110	52.8	232.0	64	64.0	235.0	65	227.6	272.3	3.9	0.97	148,993	1.49	742	33.15	0.85
3	1.5	15	120	54.0	245.0	68	62.6	229.0	64	220.4	273.4	4.0	0.97	183,462	1.83	779	33.3	0.9
4	0.75	5	0	36.8	360.0	100	50.4	360.0	100	102.2	140.0	6.0	0.73				33.8	3.2
4	0.75	5	10	31.5	320.0	89	43.2	320.0	89	98.4	135.0	5.3	0.62	51,034	0.51	229	27.2	4.6
4	0.75	5	25	30.2	266.0	74	46.6	285.0	79	113.5	163.5	4.6	0.64	148,469	1.48	680	32.2	2
4	0.75	5	30	36.4	313.0	87	46.5	282.0	78	116.3	164.9	5.0	0.69	82,413	0.82	369	30.1	4
4	0.75	5	50	39.0	267.0	74	46.0	276.0	77	146.1	166.7	4.5	0.71	267,181	2.67	1,185	31.7	3.9
4	0.75	5	65	41.8	306.0	85	49.4	273.0	76	136.6	181.0	4.8	0.76	210,784	2.11	976	31.4	3.4

4	0.75	5	80	43.8	303.0	84	51.6	267.0	74	144.6	193.3	4.8	0.80	191,985	1.92	917	31.2	2.2
4	0.75	5	95	47.5	304.0	84	52.7	250.0	69	156.3	210.8	4.6	0.84	156,182	1.56	729	32.6	1.4
4	0.75	5	110	46.0	292.0	81	52.0	265.0	74	157.5	196.2	4.6	0.82	141,583	1.42	698	32.9	1.6
4	0.75	5	120	49.0	282.0	78	53.0	254.0	71	173.8	208.7	4.5	0.85	111,951	1.12	485	33.2	1.8
5	1.5	60	0	38.8	360.0	100	50.2	360.0	100	107.8	139.4	6.0	0.74				34.1	3.3
5	1.5	60	10	60.0	320.0	89	50.2	320.0	89	187.5	156.9	5.3	0.92	67,288	0.67	302	27.1	4.4
5	1.5	60	25	42.2	316.0	88	50.3	310.0	86	133.5	162.3	5.2	0.77	144,612	1.45	646	32.1	0.6
5	1.5	60	30	42.8	316.0	88	49.7	289.0	80	135.4	172.0	5.0	0.77	29,191	0.29	132	30.0	2.8
5	1.5	60	50	49.4	308.0	86	58.2	270.0	75	160.4	215.6	4.8	0.90	297,999	2.98	1,316	31.2	2.6
5	1.5	60	65	50.6	296.0	82	54.4	258.0	72	170.9	210.9	4.6	0.88	266,189	2.66	1,240	30.8	2.4
5	1.5	60	80	55.0	258.0	72	64.0	262.0	73	213.2	244.3	4.3	0.99	189,403	1.89	894	30.6	0.9
5	1.5	60	95	61.8	275.0	76	74.2	265.0	74	224.7	280.0	4.5	1.13	154,326	1.54	756	31.9	0.6
5	1.5	60	110	63.8	268.0	74	50.8	261.0	73	238.1	194.6	4.4	0.96	198,658	1.99	964	32.3	0.85
5	1.5	60	120	66.9	263.0	73	53.6	268.0	74	254.4	200.0	4.4	1.00	189,143	1.89	817	32.7	1.1
6	0.75	15	0	44.2	360.0	100	34.8	360.0	100	122.8	96.7	6.0	0.66				33.9	3.4
6	0.75	15	10	37.8	320.0	89	40.0	320.0	89	118.1	125.0	5.3	0.65	38,853	0.39	175	27.3	4.3
6	0.75	15	25	39.8	281.0	78	36.7	275.0	76	141.6	133.5	4.6	0.64	154,556	1.55	702	32.2	0.9
6	0.75	15	30	35.2	320.0	89	38.2	300.0	83	110.0	127.3	5.2	0.61	61,060	0.61	273	30.1	2.6
6	0.75	15	50	36.2	313.0	87	37.4	274.0	76	115.7	136.5	4.9	0.61	257,125	2.57	1,138	31.3	3
6	0.75	15	65	39.2	295.0	82	41.2	264.0	73	132.9	156.1	4.7	0.67	215,855	2.16	1,015	31.1	3
6	0.75	15	80	41.8	301.0	84	41.4	250.0	69	138.9	165.6	4.6	0.69	183,634	1.84	884	31.0	2
6	0.75	15	95	42.4	281.0	78	42.8	243.0	68	150.9	176.1	4.4	0.71	150,210	1.50	714	32.4	1.1
6	0.75	15	110	44.0	285.0	79	44.8	249.0	69	154.4	179.9	4.5	0.74	160,716	1.61	783	32.6	1.55
6	0.75	15	120	44.0	282.0	78	44.5	250.0	69	156.0	178.0	4.4	0.74	125,471	1.25	542	32.7	2
7	0.75	5	0	25.1	360.0	100	36.0	360.0	100	69.7	100.0	6.0	0.51				34.1	2.8
7	0.75	5	10	31.4	320.0	89	31.0	320.0	89	98.1	96.9	5.3	0.52	24,977	0.25	110	27.6	4.2
7	0.75	5	25	32.3	380.0	106	35.8	289.0	80	85.0	123.9	5.6	0.57	132,336	1.32	600	32.6	1.8
7	0.75	5	30	27.8	320.0	89	35.5	277.0	77	86.9	128.2	5.0	0.53	55,284	0.55	249	30.9	2.3
7	0.75	5	50	28.8	296.0	82	40.0	274.0	76	97.3	146.0	4.8	0.57	226,864	2.27	1,000	31.7	3.3
7	0.75	5	65	29.4	283.0	79	39.2	270.0	75	103.9	145.2	4.6	0.57	173,043	1.73	805	31.3	3.4
7	0.75	5	80	28.8	216.0	60	42.6	252.0	70	133.3	169.0	3.9	0.60	146,988	1.47	699	30.8	2.1
7	0.75	5	95	32.2	269.0	75	43.9	257.0	71	119.7	170.8	4.4	0.63	149,492	1.49	704	32.3	1.5
7	0.75	5	110	31.8	261.0	73	44.4	248.0	69	121.8	179.0	4.2	0.64	90,433	0.90	441	32.8	1.85
7	0.75	5	120	32.5	243.0	68	46.6	251.0	70	133.7	185.7	4.1	0.66	87,756	0.88	385	33.3	2.2
8	1.5	60	0	26.2	360.0	100	38.4	360.0	100	72.8	106.7	6.0	0.54				34.2	2.2
8	1.5	60	10	36.0	320.0	89	39.0	320.0	89	112.5	121.9	5.3	0.63	33,414	0.33	150	27.8	4.7
8	1.5	60	25	39.7	406.0	113	39.9	308.0	86	97.8	129.5	6.0	0.66	122,149	1.22	556	32.6	1
8	1.5	60	30	31.3	320.0	89	38.8	298.0	83	97.8	130.2	5.2	0.58	73,788	0.74	332	30.5	1.9
8	1.5	60	50	35.4	383.0	106	46.4	280.0	78	92.4	165.7	5.5	0.68	279,030	2.79	1,224	31.6	2.3

(Continued)

Appendix B1. (*Continued*)

Hapa No.	Feeding Rate	Hap-Exch	S-Day	FTWT	FNO	FSURV	MTWT	MNO	MSURV	FMWT	MMWT	DENSITY	BIOMASS	SEEDNO1	SEEDNO2	SEEDWT	TEMP3	DO6
8	1.5	60	65	37.2	278.0	77	52.2	312.0	87	133.8	167.3	4.9	0.75	229,266	2.29	1,070	31.4	2.7
8	1.5	60	80	41.0	275.0	76	56.2	322.0	89	149.1	174.5	5.0	0.81	254,542	2.55	1,194	30.8	1.1
8	1.5	60	95	44.2	277.0	77	61.0	280.0	78	159.6	217.9	4.6	0.88	163,251	1.63	790	33.2	0.6
8	1.5	60	110	47.8	279.0	78	64.2	284.0	79	171.3	226.1	4.7	0.93	224,195	2.24	1,099	33.1	0.95
8	1.5	60	120	48.8	252.0	70	63.0	257.0	71	193.7	245.1	4.2	0.93	208,882	2.09	901	33.0	1.3
9	1.5	15	0	27.7	360.0	100	43.8	360.0	100	76.9	121.7	6.0	0.60				34.3	2.3
9	1.5	15	10	32.0	320.0	89	34.6	320.0	89	100.0	108.1	5.3	0.56	29,679	0.30	131	28.2	4.1
9	1.5	15	25	31.5	352.0	98	38.8	258.0	72	89.5	150.4	5.1	0.59	113,711	1.14	516	32.7	0.9
9	1.5	15	30	33.8	319.0	89	41.0	273.0	76	106.0	150.2	4.9	0.62	75,477	0.75	338	30.8	1.6
9	1.5	15	50	36.6	270.0	75	45.2	230.0	64	135.6	196.5	4.2	0.68	260,420	2.60	1,153	32.1	2.2
9	1.5	15	65	40.6	291.0	81	43.2	234.0	65	139.5	184.6	4.4	0.70	264,030	2.64	1,221	31.4	2.1
9	1.5	15	80	42.2	291.0	81	47.4	224.0	62	145.0	211.6	4.3	0.75	223,550	2.24	1,066	31.1	1.8
9	1.5	15	95	48.0	279.0	78	52.0	229.0	64	172.0	227.1	4.2	0.83	191,233	1.91	911	32.6	0.8
9	1.5	15	110	48.0	268.0	74	52.4	225.0	63	179.1	232.9	4.1	0.84	205,808	2.06	993	32.9	1.15
9	1.5	15	120	48.0	265.0	74	57.5	225.0	63	181.1	255.6	4.1	0.88	169,606	1.70	725	33.1	1.5
10	0.75	60	0	27.4	360.0	100	45.9	360.0	100	76.1	127.5	6.0	0.61				34.3	3
10	0.75	60	10	25.0	320.0	89	40.2	320.0	89	78.1	125.6	5.3	0.54	69,759	0.70	295	28.2	5.2
10	0.75	60	25	26.0	320.0	89	45.7	362.0	101	81.3	126.2	5.7	0.60	74,986	0.75	337	33	1.2
10	0.75	60	30	27.0	315.0	88	43.2	310.0	86	85.7	139.4	5.2	0.59	37,116	0.37	175	31	2.5
10	0.75	60	50	29.2	302.0	84	47.2	324.0	90	96.7	145.7	5.2	0.64	207,533	2.08	917	32.1	2.6
10	0.75	60	65	32.0	323.0	90	50.2	299.0	83	99.1	167.9	5.2	0.69	169,847	1.70	778	32.1	1.5
10	0.75	60	80	33.2	306.0	85	49.1	307.0	85	108.5	159.9	5.1	0.69	188,835	1.89	881	31.4	1.3
10	0.75	60	95	34.2	304.0	84	54.8	284.0	79	112.5	193.0	4.9	0.74	153,373	1.53	746	33.1	1
10	0.75	60	110	33.8	277.0	77	54.4	325.0	90	122.0	167.4	5.0	0.74	149,926	1.50	736	33.15	1.25
10	0.75	60	120	34.0	329.0	91	57.0	259.0	72	103.3	220.1	4.9	0.76	128,682	1.29	574	33.2	1.5
11	1.5	5	0	28.7	360.0	100	49.2	360.0	100	79.7	136.7	6.0	0.65				34.6	3.4
11	1.5	5	10	27.8	320.0	89	45.1	320.0	89	86.9	140.9	5.3	0.61	45,036	0.45	201	27.9	5.7
11	1.5	5	25	24.5	285.0	79	49.2	300.0	83	86.0	164.0	4.9	0.61	136,967	1.37	622	32.6	1.5
11	1.5	5	30	28.7	316.0	88	46.2	289.0	80	90.8	159.9	5.0	0.62	53,322	0.53	239	30.8	2.1
11	1.5	5	50	34.8	255.0	71	56.0	260.0	72	136.5	215.4	4.3	0.76	265,550	2.66	1,176	31.9	2.8
11	1.5	5	65	39.4	281.0	78	60.6	309.0	86	140.2	196.1	4.9	0.83	194,572	1.95	911	31.5	2.6
11	1.5	5	80	40.4	275.0	76	65.8	280.0	78	146.9	235.0	4.6	0.89	200,204	2.00	956	31.1	2
11	1.5	5	95	43.9	274.0	76	70.7	275.0	76	160.2	257.1	4.6	0.96	208,951	2.09	992	32.5	1.8
11	1.5	5	110	45.8	245.0	68	71.2	273.0	76	186.9	260.8	4.3	0.98	173,238	1.73	851	32.8	2
11	1.5	5	120	49.0	209.0	58	75.0	273.0	76	234.4	274.7	4.0	1.03	142,809	1.43	625	33.1	2.2

12	0.75	15	0	30.9	360.0	100	38.1	360.0	100	85.8	105.8	6.0	0.58	46,700	0.47	209	34.6	3.7
12	0.75	15	10	27.0	320.0	89	35.0	320.0	89	84.4	109.4	5.3	0.52	149,128	1.49	677	28	5.6
12	0.75	15	25	30.7	361.0	100	29.8	270.0	75	85.0	110.4	5.3	0.50	72,621	0.73	323	32.6	1.6
12	0.75	15	30	28.7	316.0	88	33.4	291.0	81	90.8	114.8	5.1	0.52	271,202	2.71	1,196	30.5	2.2
12	0.75	15	50	32.8	303.0	84	37.4	300.0	83	108.3	124.7	5.0	0.59	208,457	2.08	982	31.8	3
12	0.75	15	65	21.8	295.0	82	38.0	277.0	77	73.9	137.2	4.8	0.50	177,020	1.77	844	31.7	3.1
12	0.75	15	80	30.4	243.0	68	39.2	241.0	67	125.1	162.7	4.0	0.58	193,883	1.94	923	31.1	2.3
12	0.75	15	95	32.8	266.0	74	44.1	275.0	76	123.3	160.4	4.5	0.64	169,532	1.70	828	33	1.7
12	0.75	15	110	36.0	276.0	77	43.0	267.0	74	130.4	161.0	4.5	0.66	133,330	1.33	591	33.15	2
12	0.75	15	120	36.0	273.0	76	43.5	256.0	71	131.9	169.9	4.4	0.66				33.3	2.3
13	0.75	60	0	25.3	360.0	100	30.0	360.0	100	70.3	83.3	6.0	0.46	17,680	0.18	79	34.7	4
13	0.75	60	10	26.0	320.0	89	37.0	320.0	89	81.3	115.6	5.3	0.53	91,731	0.92	413	28	5.2
13	0.75	60	25	22.9	315.0	88	33.3	330.0	92	72.7	100.9	5.4	0.47	45,620	0.46	204	32.5	1.6
13	0.75	60	30	23.8	315.0	88	30.7	331.0	92	75.6	92.7	5.4	0.45	215,398	2.15	962	30.5	2.5
13	0.75	60	50	26.0	300.0	83	35.2	310.0	86	86.7	113.5	5.1	0.51	170,401	1.70	783	31.7	3.2
13	0.75	60	65	28.4	289.0	80	36.8	298.0	83	98.3	123.5	4.9	0.54	149,419	1.49	714	31.4	3.3
13	0.75	60	80	29.0	284.0	79	36.6	289.0	80	102.1	126.6	4.8	0.55	145,449	1.45	699	31.2	1.5
13	0.75	60	95	30.2	281.0	78	37.2	315.0	88	107.5	118.1	5.0	0.56	165,448	1.65	798	32.6	1.3
13	0.75	60	110	31.0	267.0	74	49.0	290.0	81	116.1	169.0	4.6	0.67	136,543	1.37	595	32.95	1.7
13	0.75	60	120	32.0	248.0	69	41.0	284.0	79	129.0	144.4	4.4	0.61				33.3	2.1
14	1.5	5	0	29.0	360.0	100	31.5	360.0	100	80.6	87.5	6.0	0.50	29,805	0.30	134	34.6	3.9
14	1.5	5	10	23.0	320.0	89	30.2	320.0	89	71.9	94.4	5.3	0.44	114,288	1.14	512	27.5	5.4
14	1.5	5	25	24.2	284.0	79	35.5	298.0	83	85.2	119.1	4.9	0.50	68,667	0.69	309	32.5	2.3
14	1.5	5	30	28.0	315.0	88	38.4	281.0	78	88.9	136.7	5.0	0.55	256,017	2.56	1,126	30.6	3.6
14	1.5	5	50	31.2	303.0	84	45.0	337.0	94	103.0	133.5	5.3	0.64	207,459	2.07	984	31.9	3.6
14	1.5	5	65	32.8	287.0	80	50.0	327.0	91	114.3	152.9	5.1	0.69	200,591	2.01	954	31.7	3.6
14	1.5	5	80	35.0	271.0	75	53.4	312.0	87	129.2	171.2	4.9	0.74	195,675	1.96	945	31.2	2.2
14	1.5	5	95	37.3	274.0	76	57.0	300.0	83	136.1	190.0	4.8	0.79	191,362	1.91	932	33.2	2
14	1.5	5	110	39.0	230.0	64	59.6	310.0	86	169.6	192.3	4.5	0.82	117,364	1.17	516	33.25	2.25
14	1.5	5	120	40.0	238.0	66	61.4	303.0	84	168.1	202.6	4.5	0.85				33.3	2.5
15	1.5	15	0	24.8	360.0	100	30.4	360.0	100	68.9	84.4	6.0	0.46	21,168	0.21	94	34.4	3.7
15	1.5	15	10	26.0	320.0	89	32.2	320.0	89	81.3	100.6	5.3	0.49	118,950	1.19	536	27.3	5.3
15	1.5	15	25	28.5	315.0	88	33.5	306.0	85	90.5	109.5	5.2	0.52	52,173	0.52	233	32.4	1.7
15	1.5	15	30	28.3	314.0	87	35.5	294.0	82	90.1	120.7	5.1	0.53	277,292	2.77	1,229	30.2	2.8
15	1.5	15	50	37.6	274.0	76	41.2	226.0	63	137.2	182.3	4.2	0.66	237,798	2.38	1,116	31.8	2.8
15	1.5	15	65	34.2	291.0	81	47.0	306.0	85	117.5	153.6	5.0	0.68	239,527	2.40	1,135	31.3	2.3
15	1.5	15	80	39.0	283.0	79	48.6	285.0	79	137.8	170.5	4.7	0.73	186,026	1.86	904	31.2	2
15	1.5	15	95	42.0	296.0	82	53.2	260.0	72	141.9	204.6	4.6	0.79	197,113	1.97	974	33.2	1.5
15	1.5	15	110	40.0	256.0	71	58.8	315.0	88	156.3	186.7	4.8	0.82	170,925	1.71	728	33.3	1.75
15	1.5	15	120	58.5	290.0	81	44.5	244.0	68	201.7	182.4	4.5	0.86				33.4	2
16	1.5	60	0	25.9	360.0	100	27.5	360.0	100	71.9	76.4	6.0	0.45				34.2	3.9

(Continued)

Appendix B1. (*Continued*)

Hapa No.	Feeding Rate	Hap-Exch	S-Day	FTWT	FNO	FSURV	MTWT	MNO	MSURV	FMWT	MMWT	DENSITY	BIOMASS	SEEDNO1	SEEDNO2	SEEDWT	TEMP3	DO6
16	1.5	60	10	37.0	320.0	89	45.0	320.0	89	115.6	140.6	5.3	0.68	24,769	0.25	110	27.4	5.8
16	1.5	60	25	40.0	450.0	125	50.0	463.0	129	88.9	108.0	7.6	0.75	133,371	1.33	598	32.5	1.4
16	1.5	60	30	36.5	315.0	88	36.0	315.0	88	115.9	114.3	5.3	0.60	60,729	0.61	272	30.1	2.8
16	1.5	60	50	44.2	376.0	104	42.2	317.0	88	117.6	133.1	5.8	0.72	326,189	3.26	1,436	31.6	2.7
16	1.5	60	65	36.0	290.0	81	34.6	333.0	93	124.1	103.9	5.2	0.59	224,431	2.24	1,031	31.5	2.6
16	1.5	60	80	38.4	289.0	80	48.0	254.0	71	132.9	189.0	4.5	0.72	189,816	1.90	886	31.1	1.5
16	1.5	60	95	43.0	290.0	81	56.6	297.0	83	148.3	190.6	4.9	0.83	143,552	1.44	685	32.8	1
16	1.5	60	110	43.0	268.0	74	67.0	303.0	84	160.4	221.1	4.8	0.92	201,321	2.01	991	33.1	1.4
16	1.5	60	120	48.7	277.0	77	58.5	282.0	78	175.8	207.4	4.7	0.89	162,956	1.63	696	33.4	1.8
17	0.75	5	0	23.3	360.0	100	28.5	360.0	100	64.7	79.2	6.0	0.43				33.8	4.1
17	0.75	5	10	32.1	320.0	89	30.1	320.0	89	100.3	94.1	5.3	0.52	22,533	0.23	102	27.3	5.6
17	0.75	5	25	24.1	290.0	81	33.3	309.0	86	83.1	107.8	5.0	0.48	118,166	1.18	530	32.5	2.3
17	0.75	5	30	24.5	307.0	85	30.8	318.0	88	79.8	96.9	5.2	0.46	58,262	0.58	261	30.2	3.8
17	0.75	5	50	24.8	278.0	77	35.2	294.0	82	89.2	119.7	4.8	0.50	210,948	2.11	939	32.2	3
17	0.75	5	65	25.2	243.0	68	38.6	287.0	80	103.7	134.5	4.4	0.53	162,942	1.63	752	31.7	2.9
17	0.75	5	80	22.2	226.0	63	34.8	258.0	72	98.2	134.9	4.0	0.48	143,931	1.44	680	31.5	2.1
17	0.75	5	95	25.0	219.0	61	38.8	216.0	60	114.2	179.6	3.6	0.53	134,980	1.35	652	33.1	1.5
17	0.75	5	110	24.8	193.0	54	37.2	247.0	69	128.5	150.6	3.7	0.52	119,276	1.19	585	33.45	1.85
17	0.75	5	120	25.0	200.0	56	41.7	236.0	66	125.0	176.7	3.6	0.56	94,260	0.94	402	33.8	2.2
18	0.75	15	0	23.5	360.0	100	26.5	360.0	100	65.3	73.6	6.0	0.42				33.6	3.9
18	0.75	15	10	32.6	320.0	89	25.2	320.0	89	101.9	78.8	5.3	0.48	18,616	0.19	82	27.2	5.3
18	0.75	15	25	22.0	303.0	84	33.6	311.0	86	72.6	108.0	5.1	0.46	93,614	0.94	421	32.5	1.6
18	0.75	15	30	34.8	312.0	87	33.4	314.0	87	111.5	106.4	5.2	0.57	50,545	0.51	227	30.3	3.1
18	0.75	15	50	37.0	251.0	70	35.2	310.0	86	147.4	113.5	4.7	0.60	221,868	2.22	974	32.1	2.6
18	0.75	15	65	30.0	289.0	80	40.0	316.0	88	103.8	126.6	5.0	0.58	168,932	1.69	797	31.6	2.9
18	0.75	15	80	29.6	229.0	64	42.0	302.0	84	129.3	139.1	4.4	0.60	168,508	1.69	804	31.4	1.8
18	0.75	15	95	29.6	261.0	73	44.0	304.0	84	113.4	144.7	4.7	0.61	173,207	1.73	842	33.3	1.1
18	0.75	15	110	32.4	256.0	71	46.8	268.0	74	126.6	174.6	4.4	0.66	174,526	1.75	853	33.65	1.45
18	0.75	15	120	33.5	280.0	78	46.3	272.0	76	119.6	170.2	4.6	0.67	137,862	1.38	597	34	1.8

Appendix B2. Descriptive statistics for each feeding level.

Feeding		FTWT	FNO	FSURV	MTWT	MNO	MSURV	FMWT	MMWT	DENSITY	BIOMSS	SEEDNO	SEEDWT
0.75	Mean	32.83	299.12	83.17	40.95	291.61	81.00	111.22	143.49	4.9211	0.6157	1.4036	594.4247
	N	90	90	90	90	90	90	90	90	90	90	81	73
	SD	7.054	38.796	10.729	6.766	35.973	10.003	25.497	33.045	0.55698	0.10092	0.65197	265.72291
	Minimum	22	193	54	25	216	60	65	74	3.60	0.42	0.18	79.00
	Maximum	56	380	106	57	362	101	180	220	6.00	0.85	3.07	982.00
	Skewness	0.689	-0.374	-0.348	0.031	0.368	0.380	0.308	0.006	0.084	0.249	-0.027	-0.433
	Kurtosis	0.259	0.216	0.192	-0.346	-0.628	-0.628	-0.286	-0.650	0.094	-0.480	-0.438	-1.066
1.50	Mean	39.77	299.59	83.23	49.84	291.31	81.00	137.29	176.38	4.9233	0.7471	1.6315	627.2222
	N	90	90	90	90	90	90	90	90	90	90	81	63
	SD	9.450	42.061	11.699	10.665	41.973	11.656	44.522	50.968	0.63459	0.15330	0.76985	296.63508
	Minimum	23	209	58	28	224	62	69	76	3.90	0.44	0.21	94.00
	Maximum	67	450	125	75	463	129	254	280	7.60	1.13	3.26	993.00
	Skewness	0.576	0.803	0.784	0.107	0.788	0.809	0.554	0.139	1.069	0.056	-0.193	-0.416
	Kurtosis	0.238	1.038	1.008	-0.363	1.863	1.932	-0.389	-0.765	2.304	-0.613	-0.729	-1.246
Total	Mean	36.30	299.36	83.20	45.40	291.46	81.00	124.26	159.94	4.9222	0.6814	1.5175	609.6176
	N	180	180	180	180	180	180	180	180	180	180	162	136
	SD	9.015	40.349	11.193	9.958	38.979	10.831	38.466	45.897	0.59538	0.14523	0.72026	279.89054
	Minimum	22	193	54	25	216	60	65	74	3.60	0.42	0.18	79.00
	Maximum	67	450	125	75	463	129	254	280	7.60	1.13	3.26	993.00
	Skewness	0.771	0.285	0.290	0.530	0.624	0.643	0.927	0.481	0.671	0.505	-0.048	-0.399
	Kurtosis	0.559	0.678	0.652	0.103	1.017	1.062	0.721	-0.178	1.486	-0.234	-0.622	-1.153

Appendix B3. Descriptive statistics for each hapa exchange interval.

hapaex		FTWT	FNO	FSURV	MTWT	MNO	MSURV	FMWT	MMWT	DENSITY	BIOMSS	SEEDNO	SEEDWT
5	Mean	34.26	286.95	79.75	46.74	289.72	80.52	122.63	164.59	4.8033	0.6758	1.3972	606.6200
	N	60	60	60	60	60	60	60	60	60	60	54	50
	SD	8.325	43.039	11.919	10.917	34.848	9.658	37.357	45.809	0.60084	0.15285	0.63581	270.33063
	Minimum	22	193	54	29	216	60	65	79	3.60	0.43	0.23	102.00
	Maximum	56	380	106	75	360	100	234	275	6.00	1.03	2.67	992.00
	Skewness	0.588	-0.008	0.016	0.505	0.493	0.507	0.804	0.330	0.286	0.465	-0.009	-0.232
	Kurtosis	-0.414	-0.286	-0.281	-0.050	-0.318	-0.316	0.266	-0.240	-0.097	-0.577	-0.742	-1.099
15	Mean	36.37	298.17	82.88	42.55	281.68	78.25	124.89	156.65	4.8333	0.6580	1.5872	621.6744
	N	60	60	60	60	60	60	60	60	60	60	54	43
	SD	7.834	35.507	9.831	8.699	40.406	11.205	35.787	48.698	0.56409	0.12686	0.74580	294.21815
	Minimum	22	229	64	25	224	62	65	74	3.90	0.42	0.19	82.00
	Maximum	59	376	104	64	360	100	228	273	6.00	0.97	3.15	993.00
	Skewness	0.511	0.336	0.323	0.501	0.385	0.402	0.733	0.651	0.538	0.513	-0.258	-.513
	Kurtosis	0.250	-0.412	-0.442	0.113	-0.738	-0.746	0.612	0.015	-0.323	-0.014	-0.659	-1.167
60	Mean	38.27	312.95	86.97	46.90	302.98	84.23	125.25	158.57	5.1300	0.7103	1.5681	601.0465
	N	60	60	60	60	60	60	60	60	60	60	54	43
	SD	10.379	38.474	10.712	9.674	39.080	10.887	42.529	43.430	0.57440	0.15196	0.77014	282.39250
	Minimum	23	248	69	28	244	68	70	76	4.20	.45	0.18	79.00
	Maximum	67	450	125	74	463	129	254	280	7.60	1.13	3.26	991.00
	Skewness	0.837	0.992	0.968	0.432	1.161	1.178	1.133	0.480	1.490	0.449	-0.018	-0.488
	Kurtosis	0.433	1.790	1.737	0.101	3.367	3.444	1.046	-0.104	4.511	-0.154	-0.520	-1.183
Total	Mean	36.30	299.36	83.20	45.40	291.46	81.00	124.26	159.94	4.9222	0.6814	1.5175	609.6176
	N	180	180	180	180	180	180	180	180	180	180	162	136
	SD	9.015	40.349	11.193	9.958	38.979	10.831	38.466	45.897	0.59538	0.14523	0.72026	279.89054
	Minimum	22	193	54	25	216	60	65	74	3.60	0.42	0.18	79.00
	Maximum	67	450	125	75	463	129	254	280	7.60	1.13	3.26	993.00
	Skewness	0.771	0.285	0.290	0.530	0.624	0.643	0.927	0.481	0.671	0.505	-0.048	-0.399
	Kurtosis	0.559	0.678	0.652	0.103	1.017	1.062	0.721	-0.178	1.486	-0.234	-0.622	-1.153

Appendix B4. Outcome of MANOVA: General Linear Model.

Source	Dependent Variable	Type III SS	Df	MS	F	Sig.	Partial Eta Squared
			Tests of Between-Subjects Effects				
Corrected Model	FTWT	9605.043[a]	51	188.334	5.636	0.000	0.774
	FNO	135738.325[b]	51	2661.536	3.875	0.000	0.702
	FSURV	10576.187[c]	51	207.376	3.997	0.000	0.708
	MTWT	10960.271[d]	51	214.907	5.315	0.000	0.763
	MNO	106117.317[e]	51	2080.732	4.130	0.000	0.715
	MSURV	8190.310[f]	51	160.594	4.118	0.000	0.714
	FMWT	189327.789[g]	51	3712.310	10.888	0.000	0.869
	MMWT	239311.664[h]	51	4692.386	8.149	0.000	0.832
	DENSITY	28.441[i]	51	0.558	5.926	0.000	0.783
	BIOMSS	2.574[j]	51	0.050	7.588	0.000	0.822
	SEEDNO	43.447[k]	51	0.852	21.430	0.000	0.929
	SEEDWT	9879163.678[l]	51	193709.092	23.360	0.000	0.934
Intercept	FTWT	1918.231	1	1918.231	57.405	0.000	0.406
	FNO	2413.854	1	2413.854	3.514	0.064	0.040
	FSURV	170.989	1	170.989	3.296	0.073	0.038
	MTWT	347.979	1	347.979	8.605	0.004	0.093
	MNO	922.972	1	922.972	1.832	0.180	0.021
	MSURV	76.365	1	76.365	1.958	0.165	0.023
	FMWT	20847.567	1	20847.567	61.147	0.000	0.421
	MMWT	11193.646	1	11193.646	19.440	0.000	0.188
	DENSITY	0.019	1	0.019	0.201	0.655	0.002
	BIOMSS	0.266	1	0.266	39.961	0.000	0.322
	SEEDNO	0.261	1	0.261	6.557	0.012	0.072
	SEEDWT	52598.681	1	52598.681	6.343	0.014	0.070
TEMP3	FTWT	1497.192	1	1497.192	44.805	0.000	0.348
	FNO	309.717	1	309.717	0.451	0.504	0.005
	FSURV	18.592	1	18.592	0.358	0.551	0.004
	MTWT	169.013	1	169.013	4.180	0.044	0.047
	MNO	3276.381	1	3276.381	6.504	0.013	0.072
	MSURV	262.748	1	262.748	6.737	0.011	0.074
	FMWT	16005.646	1	16005.646	46.945	0.000	0.359
	MMWT	6930.955	1	6930.955	12.037	0.001	0.125
	DENSITY	0.121	1	0.121	1.291	0.259	0.015
	BIOMSS	0.181	1	0.181	27.268	0.000	0.245
	SEEDNO	0.125	1	0.125	3.143	0.080	0.036
	SEEDWT	24832.290	1	24832.290	2.995	0.087	0.034
DO6	FTWT	811.662	1	811.662	24.290	0.000	0.224
	FNO	127.755	1	127.755	0.186	0.667	0.002
	FSURV	9.139	1	9.139	0.176	0.676	0.002
	MTWT	240.954	1	240.954	5.959	0.017	0.066
	MNO	6247.356	1	6247.356	12.401	0.001	0.129
	MSURV	482.540	1	482.540	12.373	0.001	0.128
	FMWT	9390.156	1	9390.156	27.542	0.000	0.247
	MMWT	10934.904	1	10934.904	18.991	0.000	0.184
	DENSITY	0.330	1	0.330	3.504	0.065	0.040

(Continued)

Appendix B4. (*Continued*)

		Tests of Between-Subjects Effects					
Source	Dependent Variable	Type III SS	Df	MS	F	Sig.	Partial Eta Squared
Feeding	BIOMSS	0.134	1	0.134	20.079	0.000	0.193
	SEEDNO	0.023	1	0.023	0.581	0.448	0.007
	SEEDWT	3101.457	1	3101.457	0.374	0.542	0.004
	FTWT	867.523	1	867.523	25.961	0.000	0.236
	FNO	20.695	1	20.695	0.030	0.863	0.000
	FSURV	2.210	1	2.210	0.043	0.837	0.001
	MTWT	1691.587	1	1691.587	41.832	0.000	0.332
	MNO	130.485	1	130.485	0.259	0.612	0.003
	MSURV	13.743	1	13.743	0.352	0.554	0.004
	FMWT	13684.607	1	13684.607	40.137	0.000	0.323
	MMWT	21503.208	1	21503.208	37.346	0.000	0.308
	DENSITY	0.004	1	0.004	0.042	0.838	0.000
hapaex	BIOMSS	0.345	1	0.345	51.806	0.000	0.381
	SEEDNO	0.420	1	0.420	10.571	0.002	0.112
	SEEDWT	93241.857	1	93241.857	11.244	0.001	0.118
	FTWT	0.413	2	0.206	0.006	0.994	0.000
	FNO	8223.716	2	4111.858	5.987	0.004	0.125
	FSURV	637.907	2	318.954	6.148	0.003	0.128
	MTWT	470.162	2	235.081	5.813	0.004	0.122
	MNO	15121.547	2	7560.773	15.008	0.000	0.263
	MSURV	1182.175	2	591.087	15.156	0.000	0.265
	FMWT	1713.662	2	856.831	2.513	0.087	0.056
	MMWT	6529.846	2	3264.923	5.670	0.005	0.119
	DENSITY	3.080	2	1.540	16.365	0.000	0.280
	BIOMSS	0.035	2	0.018	2.664	0.076	0.060
	SEEDNO	0.216	2	0.108	2.718	0.072	0.061
	SEEDWT	48895.132	2	24447.566	2.948	0.058	0.066
Sampling	FTWT	3847.251	8	480.906	14.392	0.000	0.578
	FNO	64185.596	8	8023.199	11.681	0.000	0.527
	FSURV	5031.138	8	628.892	12.123	0.000	0.536
	MTWT	2793.002	8	349.125	8.634	0.000	0.451
	MNO	31954.766	8	3994.346	7.929	0.000	0.430
	MSURV	2441.691	8	305.211	7.826	0.000	0.427
	FMWT	88423.297	8	11052.912	32.419	0.000	0.755
	MMWT	81172.142	8	10146.518	17.622	0.000	0.627
	DENSITY	12.744	8	1.593	16.928	0.000	0.617
	BIOMSS	0.855	8	0.107	16.067	0.000	0.605
	SEEDNO	15.982	8	1.998	50.255	0.000	0.827
	SEEDWT	3840533.723	8	480066.715	57.892	0.000	0.846
Feeding* hapaex	FTWT	147.038	2	73.519	2.200	0.117	0.050
	FNO	274.608	2	137.304	0.200	0.819	0.005
	FSURV	18.775	2	9.387	0.181	0.835	0.004
	MTWT	159.873	2	79.936	1.977	0.145	0.045
	MNO	2917.741	2	1458.871	2.896	0.061	0.065
	MSURV	204.767	2	102.384	2.625	0.078	0.059
	FMWT	2150.389	2	1075.195	3.154	0.048	0.070
	MMWT	1014.999	2	507.500	0.881	0.418	0.021
	DENSITY	0.210	2	0.105	1.114	0.333	0.026

Appendix B4. (*Continued*)

			Tests of Between-Subjects Effects				
Source	Dependent Variable	Type III SS	Df	MS	F	Sig.	Partial Eta Squared
Feeding* Sampling	BIOMSS	0.019	2	0.010	1.462	0.238	0.034
	SEEDNO	0.048	2	0.024	0.598	0.552	0.014
	SEEDWT	11782.911	2	5891.456	0.710	0.494	0.017
	FTWT	572.727	8	71.591	2.142	0.040	0.169
	FNO	4730.290	8	591.286	0.861	0.553	0.076
	FSURV	363.443	8	45.430	0.876	0.540	0.077
	MTWT	742.557	8	92.820	2.295	0.028	0.179
	MNO	2532.468	8	316.559	0.628	0.752	0.056
	MSURV	183.659	8	22.957	0.589	0.785	0.053
	FMWT	14880.478	8	1860.060	5.456	0.000	0.342
	MMWT	12351.808	8	1543.976	2.681	0.011	0.203
	DENSITY	0.614	8	0.077	0.815	0.591	0.072
	BIOMSS	0.178	8	0.022	3.347	0.002	0.242
	SEEDNO	0.501	8	0.063	1.575	0.145	0.130
	SEEDWT	108732.641	8	13591.580	1.639	0.126	0.135
hapaex* Sampling	FTWT	467.439	16	29.215	0.874	0.600	0.143
	FNO	9447.074	16	590.442	0.860	0.616	0.141
	FSURV	715.300	16	44.706	0.862	0.614	0.141
	MTWT	313.063	16	19.566	0.484	0.949	0.084
	MNO	10968.967	16	685.560	1.361	0.182	0.206
	MSURV	894.911	16	55.932	1.434	0.146	0.215
	FMWT	7608.158	16	475.510	1.395	0.164	0.210
	MMWT	9214.704	16	575.919	1.000	0.465	0.160
	DENSITY	2.009	16	0.126	1.334	0.196	0.203
	BIOMSS	0.091	16	0.006	0.855	0.621	0.140
	SEEDNO	1.795	16	0.112	2.823	0.001	0.350
	SEEDWT	358499.112	16	22406.195	2.702	0.002	0.340
Feeding* hapaex* Sampling	FTWT	263.851	12	21.988	0.658	0.786	0.086
	FNO	6862.009	12	571.834	0.833	0.617	0.106
	FSURV	532.704	12	44.392	0.856	0.594	0.109
	MTWT	149.645	12	12.470	0.308	0.986	0.042
	MNO	5852.889	12	487.741	0.968	0.485	0.122
	MSURV	443.738	12	36.978	0.948	0.504	0.119
	FMWT	1389.351	12	115.779	0.340	0.979	0.046
	MMWT	3202.589	12	266.882	0.464	0.930	0.062
	DENSITY	1.344	12	0.112	1.190	0.304	0.145
	BIOMSS	0.040	12	0.003	0.496	0.912	0.066
	SEEDNO	0.482	12	0.040	1.011	0.446	0.126
	SEEDWT	100924.737	12	8410.395	1.014	0.444	0.127
Error	FTWT	2806.932	84	33.416			
	FNO	57694.146	84	686.835			
	FSURV	4357.696	84	51.877			
	MTWT	3396.753	84	40.438			

(*Continued*)

Appendix B4. (*Continued*)

Source	Dependent Variable	Type III SS	Df	MS	F	Sig.	Partial Eta Squared
	MNO	42316.558	84	503.769			
	MSURV	3276.036	84	39.000			
	FMWT	28639.251	84	340.943			
	MMWT	48366.390	84	575.790			
	DENSITY	7.905	84	0.094			
	BIOMSS	0.559	84	0.007			
	SEEDNO	3.339	84	0.040			
	SEEDWT	696562.440	84	8292.410			
Total	FTWT	194343.340	136				
	FNO	11731010.000	136				
	FSURV	906582.000	136				
	MTWT	301903.070	136				
	MNO	11243537.000	136				
	MSURV	868637.000	136				
	FMWT	2454826.590	136				
	MMWT	3962149.680	136				
	DENSITY	3176.510	136				
	BIOMSS	67.580	136				
	SEEDNO	282.302	136				
	SEEDWT	61117906.000	136				
Corrected Total	FTWT	12411.975	135				
	FNO	193432.471	135				
	FSURV	14933.882	135				
	MTWT	14357.024	135				
	MNO	148433.875	135				
	MSURV	11466.346	135				
	FMWT	217967.039	135				
	MMWT	287678.054	135				
	DENSITY	36.346	135				
	BIOMSS	3.133	135				
	SEEDNO	46.786	135				
	SEEDWT	10575726.118	135				

[a] $R^2 = 0.774$ (Adjusted $R^2 = 0.637$)
[b] $R^2 = 0.702$ (Adjusted $R^2 = 0.521$)
[c] $R^2 = 0.708$ (Adjusted $R^2 = 0.531$)
[d] $R^2 = 0.763$ (Adjusted $R^2 = 0.620$)
[e] $R^2 = 0.715$ (Adjusted $R^2 = 0.542$)
[f] $R^2 = 0.714$ (Adjusted $R^2 = 0.541$)
[g] $R^2 = 0.869$ (Adjusted $R^2 = 0.789$)
[h] $R^2 = 0.832$ (Adjusted $R^2 = 0.730$)
[i] $R^2 = 0.783$ (Adjusted $R^2 = 0.650$)
[j] $R^2 = 0.822$ (Adjusted $R^2 = 0.713$)
[k] $R^2 = 0.929$ (Adjusted $R^2 = 0.885$)
[l] $R^2 = 0.934$ (Adjusted $R^2 = 0.894$)

Appendix B5. Outcome of MANOVA for factor 1, i.e. feeding level (pairwise comparisons).

Dependent Variable	(*I*) Feeding	(*J*) Feeding	Mean Difference (*I* − *J*)	Std. Error	Sig.[a]	95% CI for Difference[a] Lower level	95% CI for Difference[a] Upper level
FTWT	0.75	1.50	−5.990*,b	1.135	0.000	−8.246	−3.734
	1.50	0.75	5.990*,c	1.135	0.000	3.734	8.246
FNO	0.75	1.50	0.901b	5.144	0.861	−9.329	11.131
	1.50	0.75	−0.901c	5.144	0.861	−11.131	9.329
FSURV	0.75	1.50	0.298b	1.414	0.834	−2.514	3.109
	1.50	0.75	−0.298c	1.414	0.834	−3.109	2.514
MTWT	0.75	1.50	−9.112*,b	1.248	0.000	−11.594	−6.630
	1.50	0.75	9.112*,c	1.248	0.000	6.630	11.594
MNO	0.75	1.50	−2.381b	4.406	0.590	−11.142	6.380
	1.50	0.75	2.381c	4.406	0.590	−6.380	11.142
MSURV	0.75	1.50	−0.768b	1.226	0.533	−3.205	1.670
	1.50	0.75	0.768c	1.226	0.533	−1.670	3.205
FMWT	0.75	1.50	−24.050*,b	3.624	0.000	−31.258	−16.843
	1.50	0.75	24.050*,c	3.624	0.000	16.843	31.258
MMWT	0.75	1.50	−32.225*,b	4.710	0.000	−41.591	−22.858
	1.50	0.75	32.225*,c	4.710	0.000	22.858	41.591
DENSITY	0.75	1.50	−0.012b	0.060	0.839	−0.132	0.107
	1.50	0.75	0.012c	0.060	0.839	−0.107	0.132
BIOMSS	0.75	1.50	−0.126*,b	0.016	0.000	−0.157	−0.094
	1.50	0.75	0.126*,c	0.016	0.000	0.094	0.157
SEEDNO	0.75	1.50	−0.038b	0.039	0.330	−0.116	0.039
	1.50	0.75	0.038c	0.039	0.330	−0.039	0.116
SEEDWT	0.75	1.50	−22.623b	17.874	0.209	−58.168	12.922
	1.50	0.75	22.623c	17.874	0.209	−12.922	58.168

Based on estimated marginal means
* The mean difference is significant at the 0.05 level.
[a] Adjustment for multiple comparisons: LSD (equivalent to no adjustments).
[b] An estimate of the modified population marginal mean (*J*).
[c] An estimate of the modified population marginal mean (*I*).

Appendix B6. Multivariate tests for factor 1, i.e. feeding level.

	Value	F	Hypothesis df	Error df	Sig.	Partial Eta Squared
Pillai's trace	0.584	8.544[a]	12.000	73.000	0.000	0.584
Wilks' lambda	0.416	8.544[a]	12.000	73.000	0.000	0.584
Hotelling's trace	1.404	8.544[a]	12.000	73.000	0.000	0.584
Roy's largest root	1.404	8.544[a]	12.000	73.000	0.000	0.584

Each F tests the multivariate effect of feeding. These tests are based on the linearly independent pairwise comparisons among the estimated marginal means.
[a] Exact statistic.

Appendix B7. Univariate tests for factor 1, i.e. feeding level.

Dependent Variable		SS	df	MS	F	Sig.	Partial Eta Squared
FTWT	Contrast	931.248	1	931.248	27.868	0.000	0.249
	Error	2806.932	84	33.416			
FNO	Contrast	21.070	1	21.070	0.031	0.861	0.000
	Error	57694.146	84	686.835			
FSURV	Contrast	2.300	1	2.300	0.044	0.834	0.001
	Error	4357.696	84	51.877			
MTWT	Contrast	2155.163	1	2155.163	53.296	0.000	0.388
	Error	3396.753	84	40.438			
MNO	Contrast	147.186	1	147.186	0.292	0.590	0.003
	Error	42316.558	84	503.769			
MSURV	Contrast	15.295	1	15.295	0.392	0.533	0.005
	Error	3276.036	84	39.000			
FMWT	Contrast	15012.946	1	15012.946	44.034	0.000	0.344
	Error	28639.251	84	340.943			
MMWT	Contrast	26952.554	1	26952.554	46.810	0.000	0.358
	Error	48366.390	84	575.790			
DENSITY	Contrast	0.004	1	0.004	0.042	0.839	0.000
	Error	7.905	84	0.094			
BIOMSS	Contrast	0.409	1	0.409	61.527	0.000	0.423
	Error	0.559	84	0.007			
SEEDNO	Contrast	0.038	1	0.038	0.959	0.330	0.011
	Error	3.339	84	0.040			
SEEDWT	Contrast	13283.789	1	13283.789	1.602	0.209	0.019
	Error	696562.440	84	8292.410			

The F tests the effect of feeding. This test is based on the linearly independent pairwise comparisons among the estimated marginal means.

Appendix B8. Outcome of MANOVA for factor 2, i.e. hapa exchange interval.

Dependent Variable	(*I*) hapaex	(*J*) hapaex	Mean Difference (*I − J*)	SE	Sig.[a]	95% CI for Difference[a] Lower Level	Upper Level
FTWT	5	15	0.677[b]	1.369	0.622	−2.046	3.400
		60	0.874[b]	1.441	0.546	−1.991	3.739
	15	5	−0.677[c]	1.369	0.622	−3.400	2.046
		60	0.197[b,c]	1.328	0.883	−2.444	2.838
	60	5	−0.874[c]	1.441	0.546	−3.739	1.991
		15	−0.197[b,c]	1.328	0.883	−2.838	2.444
FNO	5	15	−8.242[b]	6.208	0.188	−20.588	4.103
		60	−23.039*,[b]	6.532	0.001	−36.028	−10.051
	15	5	8.242[c]	6.208	0.188	−4.103	20.588
		60	−14.797*,[b,c]	6.021	0.016	−26.770	−2.824
	60	5	23.039*,[c]	6.532	0.001	10.051	36.028
		15	14.797*,[b,c]	6.021	0.016	2.824	26.770
FSURV	5	15	−2.358[b]	1.706	0.171	−5.751	1.035
		60	−6.430*,[b]	1.795	0.001	−10.000	−2.861
	15	5	2.358[c]	1.706	0.171	−1.035	5.751
		60	−4.073*,[b,c]	1.655	0.016	−7.363	−0.782
	60	5	6.430*,[c]	1.795	0.001	2.861	10.000
		15	4.073*,[b,c]	1.655	0.016	0.782	7.363
MTWT	5	15	5.570*,[b]	1.506	0.000	2.574	8.565
		60	1.778[b]	1.585	0.265	−1.374	4.929
	15	5	−5.570*,[c]	1.506	0.000	−8.565	−2.574
		60	−3.792*,[b,c]	1.461	0.011	−6.697	−0.887
	60	5	−1.778[c]	1.585	0.265	−4.929	1.374
		15	3.792*,[b,c]	1.461	0.011	0.887	6.697
MNO	5	15	−2.775[b]	5.317	0.603	−13.348	7.797
		60	−28.703*,[b]	5.594	0.000	−39.827	−17.579
	15	5	2.775[c]	5.317	0.603	−7.797	13.348
		60	−25.927*,[b,c]	5.156	0.000	−36.181	−15.673
	60	5	28.703*,[c]	5.594	0.000	17.579	39.827
		15	25.927*,[b,c]	5.156	0.000	15.673	36.181
MSURV	5	15	−0.732[b]	1.479	0.622	−3.674	2.210
		60	−8.001*,[b]	1.556	0.000	−11.096	−4.906
	15	v5	0.732[c]	1.479	0.622	−2.210	3.674
		60	−7.269*,[b,c]	1.435	0.000	−10.122	−4.416
	60	5	8.001*,[c]	1.556	0.000	4.906	11.096
		15	7.269*,[b,c]	1.435	0.000	4.416	10.122
FMWT	5	15	5.534[b]	4.374	0.209	−3.164	14.232
		60	12.911*,[b]	4.602	0.006	3.760	22.063
	15	5	−5.534[c]	4.374	0.209	−14.232	3.164
		60	7.377[b,c]	4.242	0.086	−1.058	15.813
	60	5	−12.911*,[c]	4.602	0.006	−22.063	−3.760
		15	−7.377[b,c]	4.242	0.086	−15.813	1.058
MMWT	5	15	18.965*,[b]	5.684	0.001	7.662	30.268
		60	22.176*,[b]	5.980	0.000	10.284	34.068
	15	5	−18.965*,[c]	5.684	0.001	−30.268	−7.662
		60	3.211[b,c]	5.513	0.562	−7.751	14.173

(Continued)

Appendix B8. (*Continued*)

Dependent Variable	(*I*) hapaex	(*J*) hapaex	Mean Difference (*I − J*)	SE	Sig.[a]	95% CI for Difference[a] Lower Level	Upper Level
	60	5	−22.176*,c	5.980	0.000	−34.068	−10.284
		15	−3.211b,c	5.513	0.562	−14.173	7.751
DENSITY	5	15	−0.093b	0.073	0.203	−0.238	0.051
		60	−0.430*,b	0.076	0.000	−0.582	−0.278
	15	5	0.093c	0.073	0.203	−0.051	0.238
		60	−0.337*,b,c	0.070	0.000	−0.477	−0.197
	60	5	0.430*,c	0.076	0.000	0.278	0.582
		15	0.337*,b,c	0.070	0.000	0.197	0.477
BIOMSS	5	15	0.053*,b	0.019	0.008	0.014	0.091
		60	0.022b	0.020	0.276	−0.018	0.063
	15	5	−0.053*,c	0.019	0.008	−0.091	−0.014
		60	−0.031b,c	0.019	0.106	−0.068	0.007
	60	5	−0.022c	0.020	0.276	−0.063	0.018
		15	0.031b,c	0.019	0.106	−0.007	0.068
SEEDNO	5	15	−0.035b	0.047	0.459	−0.129	0.059
		60	0.032b	0.050	0.523	−0.067	0.131
	15	5	0.035c	0.047	0.459	−0.059	0.129
		60	0.067b,c	0.046	0.147	−0.024	0.158
	60	5	−0.032c	0.050	0.523	−0.131	0.067
		15	−0.067b,c	0.046	0.147	−0.158	0.024
SEEDWT	5	15	−19.829b	21.571	0.361	−62.725	23.066
		60	13.439b	22.695	0.555	−31.692	58.571
	15	5	19.829c	21.571	0.361	−23.066	62.725
		60	33.269b,c	20.920	0.116	−8.333	74.871
	60	5	−13.439c	22.695	0.555	−58.571	31.692
		15	−33.269b,c	20.920	0.116	−74.871	8.333

Based on estimated marginal means.
* The mean difference is significant at the 0.05 level.
[a] Adjustment for multiple comparisons: LSD (equivalent to no adjustments).
[b] An estimate of the modified population marginal mean (*J*).
[c] An estimate of the modified population marginal mean (*I*).

Appendix B9. Multivariate tests for factor 2, i.e. hapa exchange.

	Value	F	Hypothesis df	Error df	Sig.	Partial Eta Squared
Pillai's trace	0.748	3.683	24.000	148.000	0.000	0.374
Wilks' lambda	0.386	3.712a	24.000	146.000	0.000	0.379
Hotelling's trace	1.246	3.739	24.000	144.000	0.000	0.384
Roy's largest root	0.828	5.107b	12.000	74.000	0.000	0.453

Each F tests the multivariate effect of hapaex. These tests are based on the linearly independent pairwise comparisons among the estimated marginal means.
[a] Exact statistic
[b] The statistic is an upper bound on F that yields a lower bound on the significance level.

Appendix B10. Univariate tests for factor 2, i.e. hapa exchange.

Dependent Variable		SS	df	MS	F	Sig.	Partial Eta Squared
FTWT	Contrast	13.506	2	6.753	.202	0.817	0.005
	Error	2806.932	84	33.416			
FNO	Contrast	8936.345	2	4468.173	6.505	0.002	0.134
	Error	57694.146	84	686.835			
FSURV	Contrast	693.005	2	346.502	6.679	0.002	0.137
	Error	4357.696	84	51.877			
MTWT	Contrast	603.124	2	301.562	7.457	0.001	0.151
	Error	3396.753	84	40.438			
MNO	Contrast	17185.104	2	8592.552	17.057	0.000	0.289
	Error	42316.558	84	503.769			
MSURV	Contrast	1342.879	2	671.439	17.216	0.000	0.291
	Error	3276.036	84	39.000			
FMWT	Contrast	2725.311	2	1362.655	3.997	0.022	0.087
	Error	28639.251	84	340.943			
MMWT	Contrast	9281.645	2	4640.822	8.060	0.001	0.161
	Error	48366.390	84	575.790			
DENSITY	Contrast	3.441	2	1.720	18.281	0.000	0.303
	Error	7.905	84	0.094			
BIOMSS	Contrast	0.052	2	0.026	3.877	0.025	0.085
	Error	0.559	84	0.007			
SEEDNO	Contrast	0.086	2	0.043	1.077	0.345	0.025
	Error	3.339	84	0.040			
SEEDWT	Contrast	21536.816	2	10768.408	1.299	0.278	0.030
	Error	696562.440	84	8292.410			

The F tests the effect of hapaex. This test is based on the linearly independent pairwise comparisons among the estimated marginal means.

Appendix B11. Outcome of MANOVA for factor 3, i.e. sampling days.

	Value	F	Hypothesis df	Error df	Sig.	Partial Eta Squared
Pillai's trace	3.381	4.879	96.000	640.000	0.000	0.423
Wilks' lambda	.001	10.557	96.000	502.018	0.000	0.606
Hotelling's trace	32.848	24.380	96.000	570.000	0.000	0.804
Roy's largest root	23.101	154.009[a]	12.000	80.000	0.000	0.959

Each F tests the multivariate effect of sampling. These tests are based on the linearly independent pairwise comparisons among the estimated marginal means.
[a] The statistic is an upper bound on F that yields a lower bound on the significance level.

Appendix B12. Univariate tests for factor 3, i.e. sampling day.

Dependent Variable		SS	df	MS	F	Sig.	Partial Eta Squared
FTWT	Contrast	4010.872	8	501.359	15.004	0.000	0.588
	Error	2806.932	84	33.416			
FNO	Contrast	64145.872	8	8018.234	11.674	0.000	0.526
	Error	57694.146	84	686.835			
FSURV	Contrast	5028.615	8	628.577	12.117	0.000	0.536
	Error	4357.696	84	51.877			
MTWT	Contrast	2868.618	8	358.577	8.867	0.000	0.458
	Error	3396.753	84	40.438			
MNO	Contrast	31972.972	8	3996.622	7.933	0.000	0.430
	Error	42316.558	84	503.769			
MSURV	Contrast	2443.064	8	305.383	7.830	0.000	0.427
	Error	3276.036	84	39.000			
FMWT	Contrast	90771.905	8	11346.488	33.280	0.000	0.760
	Error	28639.251	84	340.943			
MMWT	Contrast	83146.568	8	10393.321	18.051	0.000	0.632
	Error	48366.390	84	575.790			
DENSITY	Contrast	12.711	8	1.589	16.883	0.000	0.617
	Error	7.905	84	0.094			
BIOMSS	Contrast	.888	8	0.111	16.682	0.000	0.614
	Error	.559	84	0.007			
SEEDNO	Contrast	16.938	8	2.117	53.262	0.000	0.835
	Error	3.339	84	0.040			
SEEDWT	Contrast	4074788.057	8	509348.507	61.423	0.000	0.854
	Error	696562.440	84	8292.410			

The F tests the effect of sampling. This test is based on the linearly independent pairwise comparisons among the estimated marginal means.

Appendix B13. Post-hoc test or multiple comparisons for the variable SEEDWT.

	hapaex	N	Subset 1
Tukey's HSD[a,b,c]	60	43	601.0465
	5	50	606.6200
	15	43	621.6744
	Sig.		0.536
Duncan[a,b,c]	60	43	601.0465
	5	50	606.6200
	15	43	621.6744
	Sig.		0.319

Means for groups in homogeneous subsets are displayed.
Based on Type III SS.
The error term is MS(Error) = 8399.153.
[a] Uses harmonic mean sample size = 45.105.
[b] The group sizes are unequal. The harmonic mean of the group sizes is used. Type I error levels are not guaranteed.
[c] Alpha = 0.05.

Appendix B14. Post-hoc test or multiple comparisons for the variable SEEDNO.

	hapaex	N	Subset 1
Tukey's HSD[a,b,c]	60	43	1.3044
	5	50	1.3058
	15	43	1.3393
	Sig.		0.689
Duncan[a,b,c]	60	43	1.3044
	5	50	1.3058
	15	43	1.3393
	Sig.		0.442

Means for groups in homogeneous subsets are displayed.
Based on type III SS.
The error term is MS(Error) = 0.040.
[a] Uses harmonic mean sample size = 45.105.
[b] The group sizes are unequal. The harmonic mean of the group sizes is used. Type I error levels are not guaranteed.
[c] Alpha = 0.05.

Appendix B15. Post-hoc test or multiple comparisons for the variable BIOMSS.

	hapaex	N	Subset 1	Subset 2
Tukey's HSD[a,b,c]	15	43	0.6619	
	5	50	0.6872	0.6872
	60	43		0.7163
	Sig.		0.442	0.343
Duncan[a,b,c]	15	43	0.6619	
	5	50	0.6872	0.6872
	60	43		0.7163
	Sig.		0.224	0.164

Means for groups in homogeneous subsets are displayed.
Based on Type III SS.
The error term is MS(Error) = 0.010.
[a] Uses harmonic mean sample size = 45.105.
[b] The group sizes are unequal. The harmonic mean of the group sizes is used. Type I error levels are not guaranteed.
[c] Alpha = 0 .05.

Appendix B16. Post-hoc test or multiple comparisons for the variable DENSITY.

	hapaex	N	Subset 1	Subset 2
Tukey's HSD[a,b]	5	50	4.6660	
	15	43	4.7326	
	60	43		5.0395
	Sig.		0.568	1.000
Duncan[a,b]	5	50	4.6660	
	15	43	4.7326	
	60	43		5.0395
	Sig.		0.312	1.000

Means for groups in homogeneous subsets are displayed.
Based on Type III SS.
The error term is MS(Error) = 0.097.
[a] Uses harmonic mean sample size = 45.105.
[b] Alpha = 0.05.

Appendix B17. Post-hoc test or multiple comparisons for the variable FMWT.

	hapaex	N	Subset 1
Tukey's HSD[a,b]	5	50	127.57
	60	43	128.63
	15	43	128.65
	Sig.		0.976
Duncan[a,b]	5	50	127.57
	60	43	128.63
	15	43	128.65
	Sig.		0.845

Means for groups in homogeneous subsets are displayed.
Based on Type III SS.
The error term is MS(Error) = 588.801.
[a] Uses harmonic mean sample size = 45.105.
[b] Alpha = 0.05.

Appendix B18. Post-hoc test or multiple comparisons for the variable MSURV.

	hapaex	N	Subset 1	Subset 2
Tukey's HSD[a,b,c]	15	43	76.81	
	5	50	78.24	
	60	43		83.30
	Sig.		0.577	1.000
Duncan[a,b,c]	15	43	76.81	
	5	50	78.24	
	60	43		83.30
	Sig.		0.319	1.000

Means for groups in homogeneous subsets are displayed.
Based on Type III SS.
The error term is MS(Error) = 45.614.
[a] Uses harmonic mean sample size = 45.105.
[b] The group sizes are unequal. The harmonic mean of the group sizes is used. Type I error levels are not guaranteed.
[c] Alpha = 0.05.

Appendix B19. Post-hoc test or multiple comparisons for the variable FSURV.

	hapaex	N	Subset 1	Subset 2	Subset 3
Tukey's HSD[a,b]	5	50	77.48		
	15	43	81.05		
	60	43		84.95	
	Sig.		0.051	1.000	
Duncan[a,b]	5	50	77.48		
	15	43		81.05	
	60	43			84.95
	Sig.		1.000	1.000	1.000

Means for groups in homogeneous subsets are displayed.
Based on Type III SS.
The error term is MS(Error) = 50.952.
[a] Uses harmonic mean sample size = 45.105.
[b] Alpha = 0.05.

Appendix B20. Post-hoc test or multiple comparisons for the variable FNO.

	hapaex	N	Subset 1	Subset 2	Subset 3
Tukey's HSD	5	50	278.72		
	15	43	291.47		
	60	43		305.65	
Sig.			0.057	1.000	
Duncan	5	50	278.72		
	15	43		291.47	
	60	43			305.65
Sig.			1.000	1.000	1.000

Appendix B21. Post-hoc test or multiple comparisons for the variable FTWT.

	hapaex	N	Subset 1	Subset 2
Tukey HSD[a,b]	5	50	34.87	
	15	43	36.66	
	60	43	38.47	
Sig.			0.062	
Duncan[a,b]	5	50	34.87	
	15	43	36.66	36.66
	60	43		38.47
Sig.			0.257	.253

Means for groups in homogeneous subsets are displayed.
Based on Type III SS.
The error term is MS(Error) = 55.941.
[a] Uses harmonic mean sample size = 45.105.
[b] Alpha = 0.05.

Bibliography

Bhujel R C, Little D C, Hossain A (2007) Reproductive performance and the growth of pre-stunted and normal Nile tilapia (*Oreochromis niloticus*) broodfish at varying feeding rates. *Aquaculture* **273**: 71–9.

FAO (2007) State of World Fisheries and Aquaculture 2006. FAO Fisheries and Aquaculture Department, Food and Agriculture Organization of the United Nations, Rome. ISSN 1020–5489

Gomez K A, Gomez A A (1984) *Statistical Procedures for Agricultural Research* 2nd edition. John Wiley & Sons

Knud-Hansen C F (1997) Exerimental design and analysis in aquaculture. In *Dynamics of Pond Aquaculture*. CRC Press, pp 325–75

Searcy-Bernal R (1994) Statistical power and aquacultural research. *Aquaculture* **127**: 371–88

Sokal R R, Rohlf F J (1969) *Biometry: The Principles and Practices of Statistics in Biological Research*, 2nd edition. WH Freeman and Co., New York

Schefler W C (1969) *Statistics for the Biological Sciences*. Addison-Wesley Publishing Co. Library of Congress Catalog no. 72–76074

Shrestha M K, Bhujel R C (1999) A preliminary trial on culture of Nile tilapia (Oreochromis niloticus) polyculture with common carp (*Cyprinus carpio*) fed with duckweed (*Spirodela* sp.) in Nepal. *Asian Fisheries Science* **12**: 83–9.

Yamane T (1967) *Statistics: An Introductory Analysis*, 2nd edition. Harper & Row, New York, pp 886–7

Zar J H (1996) *Bio-Statistical Analysis*. Prentice-Hall

Webliography

1. http://www.statsoft.com/ The home page of StatSoft, Inc. (1984–2008); covers a wide range of statistical analysis and solutions, including nonparametric tests, data mining, etc.
2. http://www.math.niu.edu/NPAR/ Web site maintained by the Northern Illinois University, Department of Mathematical Sciences and the American Statistical Association; provides information about nonparametric tests.
3. http://www.surveysystem.com/sscalc.htm Maintained by Creative Research Systems group; contains online sample size and confidence interval calculations and information about survey designs.
4. http://www.dssresearch.com/toolkit/default.asp Maintained by DSS Research Company; provides online sample size and confidence interval calculations and statistical power analysis.
5. http://www.raosoft.com/samplesize.html Web site maintained by Raosoft, Inc.; sample size calculation and other information.
6. http://home.ubalt.edu/ntsbarsh/Business-stat/otherapplets/SampleSize.htm Maintained by Professor Hossein Arsham; provides details of online statistical analysis, including ANOVA, regression, and others.
7. http://www.physics.csbsju.edu/stats/t-test.html Online Stats Calculators provided by T.W. Kirkman (1996); wide range of calculations, such as t-tests, K-S test, ANOVA, including plotting.
8. http://galton.org/cgi-bin/searchImages/galton/search/pearson/vol3a/pages/vol3a_0082.htm Collection of Galton's quotes under "Life of Francis Galton by Karl Pearson Vol 3a: image 0082."
9. http://cygnus-group.com/CIDM/stats.html Web site for The Center for Informed Decision Making (CIDM), which helps make informed decisions about important environmental, health, and safety issues.

Index

Printed and bound by CPI Group (UK) Ltd, Croydon, CR0 4YY

16/04/2025

14658421-0004